THE ANATOMICAL EXERCISES

De Motu Cordis and De Circulatione Sanguinis

in English Translation

WILLIAM HARVEY

EDITED BY
GEOFFREY KEYNES

DOVER PUBLICATIONS, INC.
New York

Published in Canada by General Publishing Company, Ltd., 30 Lesmill Road, Don Mills, Toronto, Ontario.
Published in the United Kingdom by Constable and Company, Ltd., 3 The Lanchesters, 162–164 Fulham Palace Road, London W6 9ER.

Bibliographical Note

This Dover edition, first published in 1995, is an unaltered and unabridged republication of the work published by The Nonesuch Press, London, in 1953 under the title *The Anatomical Exercises of Dr. William Harvey, De Motu Cordis 1628: De Circulatione Sanguinis 1649: The first English text of 1653 now newly edited by Geoffrey Keynes,* in a limited edition of 1450 copies. The first English translation of the two works was published by Richard Lowndes, London, in 1653. The original Latin text of *De Motu Cordis* was published at Frankfurt-am-Main, Germany, in 1628 and the Latin text of *De Circulatione Sanguinis* was first published at Cambridge, England, and at Rotterdam, The Netherlands, in 1649.

Library of Congress Cataloging-in-Publication Data

Harvey, William, 1578–1657.
    [De motu cordis. English]
    The anatomical exercises : de motu cordis and de circulatione sanguinis, in English translation / William Harvey ; edited by Geoffrey Keynes.
        p.   cm.
    "This Dover edition, first published in 1995 is an unaltered and unabridged republication of the work published by Nonesuch Press, London, in 1953 under the title: The Anatomical Exercises of Dr. William Harvey, De Motu Cordis 1628: De Circulatione Sanguinis 1649: the first English text of 1653 now newly edited by Geoffrey Keynes, in a limited printing of 1450 copies. The first English translation was published by Richard Lowndes, London, in 1653. The original Latin text of De Motu Cordis was published at Frankfurt-on-Main, Germany, in 1628 and the Latin text of De Circulatione Sanguinis was first published at Cambridge, England and at Rotterdam, The Netherlands, in 1649"—CIP t.p. verso.
    ISBN 0-486-68827-5 (pbk.)
    1. Blood—Circulation—Early works to 1800. I. Keynes, Geoffrey, Sir, 1887– . II. Harvey, William, 1578–1657. De circulatione sanguinis. English. 1653. III. Harvey, William, 1578–1653. Anatomical exercises of Dr. William Harvey. IV. Title.
QP101.H3613   1995
612.1'3—dc20   95-17389
                                                 CIP

Manufactured in the United States of America
Dover Publications, Inc., 31 East 2nd Street, Mineola, N.Y. 11501

THE
CONTENTS

DE MOTU CORDIS
Anatomical exercises concerning the Motion of the Heart and Blood in Living Creatures

To the Most Illustrious and Invincible Monarch

CHARLS

KING OF GREAT BRITAIN,
FRANCE, AND IRELAND,
DEFENDER OF
THE FAITH.

MOST GRATIOUS KING,

THE Heart *of creatures is the foundation of life, the Prince of all, the Sun of their Microcosm, on which all vegetation does depend, from whence all vigor and strength does flow. Likewise the King is the foundation of his Kingdoms, and the Sun of his Microcosm, the* Heart *of his Commonwealth, from whence all power and mercy proceeds. I was so bold to offer to your Majesty those things which are written concerning the* Heart, *so much the rather, because (according to the custom of this age) all things human are according to the pattern of man, and most things in a King according to that of the* Heart; *Therefore the knowledge of his own* Heart *cannot be unprofitable to a King, as being a divine resemblance of his actions (so us'd they small things with great to compare). You may at least, best of Kings, being plac'd in the top of human things, at the same time contemplate the Principle of Man's Body, and the Image of your Kingly power. I therefore most humbly intreat, most gracious King, accept, according to your accustom'd bounty and clemency, these new things concerning the* Heart, *who are the new light of this age, and indeed the whole* Heart *of it, a Prince abounding in vertue and grace, to whom we acknowledge our thanks to be due, for any good that* England *receives, or any pleasure that our life enjoyes:*

Your Sacred Majesties

most devoted Servant,

WILLIAM HARVEY.

To the most Excellent and most Ornate man

D. ARGENT,

PRESIDENT of the COLLEGE
of PHYSICIANS in LONDON,
his SINGULAR FRIEND,
AND THE REST OF THE DOCTORS
AND PHYSICIANS HIS MOST
LOVING COLLEGS.

S. P. D.

I Did open many times before, worthy Mr. Doctor, my opinion concerning the motion and use of the heart and Circulation of the blood new in my lectures; but being confirm'd by ocular demonstration for nine years and more in your sight, evidenced by reasons and arguments, freed from the objections of the most learned and skilfull Anatomists, desired by some, and most earnestly required by others, we have at last set it out to open view in this little Book; which, unless it were pass'd through your hands, I could hardly hope that it would come abroad entire and safe, since I can call most of you, being worthy of credit, as witnesses of those observations from which I gather truth, or confute error, who saw many of my Dissections, and in the ocular demonstrations of these things which I here assert to the senses, were us'd to stand by and assist me. And since this only Book does affirm the blood to pass forth and return through unwonted tracts, contrary to the received way, through so many ages of years insisted upon, and evidenced by innumerable, and those most famous and learned men, I was greatly afraid to suffer this little Book, otherways perfect some years ago, either to come abroad, or go beyond Sea, lest it might seem an action too full of arrogancy, if I had not first propounded it to you, confirm'd it by ocular testimony, answer'd your doubts and objections, and gotten the President's verdict in my favour; yet I was perswaded if I could maintain what I proposed in the presence of you and our College, having been famous by so many and so great men, I needed so much the lesse to be afraid of others, and that only comfort, which for the love of the truth you did grant me, might likewise be hoped for from all who were Philosophers of the same nature. For true Philosophers, who are perfectly in love with truth and wisdom, never find themselves so wise, or full of wisdom, or so abundantly satisfied in their own knowledge, but that they give place to truth whensoever, or from whosoever it comes. Nor are they so narrow spirited to believe that ever any art or science was so absolutely and perfectly taught in all points, that there is nothing remaining to the industry and diligence of others, since very many profess that the greatest part of those things which we do know, is the least of the things which we know not. Neither do Philosophers suffer themselves to be addicted to the slavery of any man's precepts, but that they give credit to their own eyes; nor do they so swear Allegiance to Mistris Antiquity, as openly to leave, or in the sight of all to desert their friend Truth. For as they think

them credulous and idle people, who at
first sight do receive and believe all
things, so do they take them for stupid
and senseless, that will not see things
manifest to the sense, nor acknowledge
the light at mid-day; and do teach as well
to decline the records of the Scepticks
as the follies of the rabble, or the fables
of Poets. Likewise, all studious, good and
honest men, do never suffer their mind
so to be o'rwhelm'd with the passions of
indignation and envy, but that they will
patiently hear what shall be spoken in behalf of the truth, or understand any thing
which is truly demonstrated to them; nor
do they think it base to change their opinion, if truth and open demonstration so
perswade them, and not think it shamefull
to desert their errors, though they be never so antient, seeing they very well know
that all men may erre, and many things
are found out by chance, which any one
may learn of another, an old man of a
child, or an understanding man of a fool.

But my loving Collegs, I had no desire
in this Treatise to make a great volume,

and to ostentate my memory and labours and my readings, in rehearsing, tossing the works, names and opinions of the Authors and writers of Anatomy, both because I do not profess to learn and teach Anatomy from the axioms of Philosophers, but from Dissections and from the fabrick of Nature. As likewise that I do not endeavour, nor think it fit, to defraud any of the antients of the honour due to them, nor provoke any of the moderns; nor do I think it seemly to contest and strive with those that have been excellent in Anatomy, and were my teachers. Moreover I would not willingly lay an aspersion of falshood upon any that is desirous of the truth, nor blemish any man by accusing him of an error; but I follow the truth only, and have bestowed both my pains and charges to that purpose, that I might bring forth something which might be both acceptable to good men, agreeable to learned men, and profitable to literature.

Farewell most excellent Doctors,
and favour your Anatomist,
WILLIAM HARVEY.

THE INDEX of the CHAPTERS

xiv

THE PROEME.

By which is Demonstrated, that those things which are already written concerning the motion and use of the Heart and Arteries, are not firm.

IT will be worth our while, seeing we are thinking of the motion, pulse, use, action, and utility of the heart and arteries, first to unfold such things as have been published by others, to take notice of those things which have been commonly spoken and taught, that those things which have been rightly spoken may be confirmed, and those which are false both by Anatomical dissection, manifold experience, and diligent and accurate observation, may be mended.

Almost all Anatomists, Physicians, and Philosophers to this day, do affirm with *Galen*, that the use of Pulsation is the same with that of Respiration, and that they differ only in one thing, that one flows from the Animal faculty, and the other from the Vital, being alike in all other things, either as touching their utility, or manner of motion. Whence they affirm (as *Hieronymus Fabricius ab Aquapendente* in his Book of Respiration, which he has newly set out), Because that the pulse of the heart and arteries is not sufficient to fan and refrigerate, that the lungs were made about the heart. Hence it appears, that whatsoever those in former times did say concerning the Systole and the Diastole, concerning the motion of the heart and arteries, they spoke it in relation to the lungs.

But since the motion and constitution of the heart is different from that of the lungs, and the motion of the arteries different from that of the breast, it is probable that divers uses and utilities should follow, and that the pulse of the heart and the use of it, as likewise that of the arteries, should differ much from the pulse and use of the breast and lungs. For if pulse and respiration do serve for the same use, and that the arteries do receive the air into their concavities in the Diastole, as they commonly say, and that in their Systole they send out fumes through the pores of the flesh and skin; as likewise that in the space betwixt the Systole and Diastole they do contain air; and that every time they do either expell Air, or Spirits, or Fumes; what will they then answer to *Galen?* who wrote a Book, that blood was naturally contain'd in the arteries, and nothing but blood, that there is neither Spirits, nor Air, as from Reasons and Experiments in the same Book we may easily gather. And if in the Diastole the arteries are fill'd with Air which they take in, and that in a greater pulse there enters a greater quantitie of Air; it will follow, that whilst there is a great pulse if you dip your whole body into a bath of Water or Oyl, that the pulse shall either be lessen'd, or much slower, since it is a hard thing for the air to pass through the body of the bath which encompasses them, and get into the arteries, if not altogether impossible. Likewise since all the arteries, as well those which lye deeper as those which are next to the skin, are distended with the same swiftness, how can the Air so freely, so swiftly, pass through the skin, flesh, and habit of the whole body, into the depth, as it can through the skin alone? And how shall the arteries of Embryons draw the air into their concavities through their mother's belly, and the body of the womb? And how shall Whales, Dolphins, and great Fishes, and all sorts of Fishes in the bottom of the Sea, take in the air, by the swift pulse in the Systole and Diastole of their arteries, through such a great mass of water? But to say that they sup up the air implanted in the water, and do return their fumes into it, is not unlike a fiction. And if in the Systole the arteries do expell their fumes out of their concavities through the pores of the flesh and skin, why not the Spirits likewise, which they say are contain'd there too, since Spirits are much thinner than fumes? And if the arteries do receive the air both in the Systole and the Diastole, and return it, as the lungs do in respiration, why do not they do this in inflicting of a wound when an arterie is cut? In the cutting of the wind-pipe by a wound it is clear, that the air does enter and return by two contrary motions. But it is clear in the section of an arterie, that the air is thrust out with one continual motion, and

3

the air does not enter and return. If the pulse of the arteries do refrigerate the parts of the body, and cool it, as the lungs do the heart it self, how do they say that the arteries do carry the blood very full of vital Spirits into all the parts which do nourish the heat of the parts, wake it when it is asleep, and recruit it being spent? and how comes it to pass, that if you tye the arteries, the parts are not only numm'd, cold, and look pale, but at last leave off to be nourished? which happens, according to *Galen*, because they are also depriv'd of that heat which did flow from above out of the heart: Since it is clear from hence, that the arteries do rather carry heat to the parts, than cooling or refrigeration. Besides, how shall the Diastole both draw Spirits from the heart to warm the parts, and likewise draw cold from outwards? Further, although some affirm, that the lungs, arteries, and heart do serve for one and the same purpose, Yet they say that the heart is the storehouse of the Spirits, and likewise that the arteries do contain Spirits and send them abroad; but contrary to the opinion of *Columbus*, they do deny that the lungs do make any Spirits or retain them. But likewise these men affirm with *Galen* against *Erasistratus* that blood is contain'd in the arteries, and not Spirits. These opinions seem to quarrel with one another, and to refute each the other, insomuch that all are not undeservedly suspected. It is manifest that the blood is contain'd in the arteries, and that the arteries alone do carry

4

out the blood, both by the Experiment of *Galen*, as likewise by the cutting of an arterie in wounds, (which *Galen* in his Book, That Blood is Contain'd in the Arteries, affirms, and in very many places) that by a great and forcible profusion the whole mass of blood will be exhausted in the space of half an hour. The experiment of *Galen* is thus, *Bind the Arterie at both ends with a little cord, and cutting it up in length, in the middle you shall find, in that place which is comprehended betwixt the two ligatures, nothing but blood, and so does he prove that it contains only blood.* Whence we may argue likewise in the same manner; If you find the same blood in the arteries which is in the veins, being bound and cut up after the same manner, as I have often tryed in dead men and in other creatures, by the same reason we may likewise conclude, that the arteries do contain the same blood which the veins, and nothing but the same blood. Some whilst they endeavour to dissolve this difficulty, affirming that it is Arterial blood and full of Spirit, they do silently grant that it is the function of the arteries to carry the blood from the heart into the whole body, and that the arteries are full of blood. (For the blood that has Spirit is no less blood). Likewise no man does deny that the blood, as it is blood, and flowes in the veins, is imbued with Spirits. Albeit the blood in the arteries do swell with greater store of Spirits, yet those Spirits are to be thought inseparable from the blood, as those which are in the veins; and that Blood and

5

Spirit make one body, as whey and butter in milk, or heat and water in warm water, by which the arteries are fill'd, and the distribution of which body from the heart the arteries do perform, and this body is nothing else but blood. But if they say that this blood is attracted out of the heart into the arteries by the Diastole of the arteries, then they seem to presuppose that the arteries by their own distention, are fill'd with that blood, and not with the ambient air as before; but if in the Diastole, they shall together receive the blood, the air, the heat, and the cold at one time, that is improbable. Further, when they do affirm that the Diastole of the heart and arteries is at one time, and so their Systole, one of these two will be inconsistent. For how shall two bodies so nearly joyn'd together, whilst they are distended, one of them draw from the other, or when they are contracted at one time, how shall one receive any thing from the other? Over and above, it may be perchance impossible, that any body should so attract into it self, as that it should be distended, seeing to be distended is to suffer, unless it do as a spunge, returning to its own natural constitution after external constriction. It were a hard thing to feign that any such thing could be in the arteries. But I believe I can easily demonstrate, and have heretofore demonstrated, that the arteries are distended, because they are fill'd like Satchells or baggs, not because they are blown up like bladders. Yet notwithstanding *Galen's* Experiment, in his Book, That Blood is Contain'd in the Arteries, is otherwise, after this manner. He did cut the arterie being laid open in length, and into the wound he thrust a reed or a hollow pipe and stop'd the wound that the blood could not leap out. *So long* (says he) *as the arterie is thus, all of it will beat, but so soon as with a thred you have above the arteries and pipe contracted the tunicle of the arterie with a noose, and stop'd it with heed, you shall not see the arterie beat any more above the noose.* I have neither tryed this experiment of *Galen's*, nor do I think it can be tryed and the body kept alive, by reason of the preruption of the blood out of the arterie, nor can the pipe close the wound without a ligature; nor do I doubt but that the blood will stream further through the concavity of the pipe. Nevertheless *Galen* by this Experiment seems to prove, that the pulsifick faculty flows through the tunicles of the arteries from the heart, and that the arteries whilst they are distended by the pulsifick faculty are fill'd, because they are distended as bellows, not distended because they are fill'd like baggs. But the contrary is manifest, both in cutting of an arterie, and in wounds; For the blood is poured out of the arteries with a forcible leaping, sometimes farther, sometimes nigher, leaping by fits, but the leaping of it is always in the Diastole of the arterie, not in the Systole. By which it appears clearly, that the arterie is distended by the impulsion of blood. For of it self it cannot by its distention throw the blood out

7

so far, it should rather attract air into it through the wound, according to those things which are commonly spoken. Nor let the thickness of the arterial tunicles cosen us in that, that the pulsifick faculty flows from the heart by the tunicles themselves; for in some creatures arteries do differ nothing from veins, and in the most remote parts of a man, and the disseminations of the arteries, as in the brain, hand, &c. no body can distinguish an arterie from a vein, for they have both the same tunicles. Besides in an Aneurism, which is begot by the arrosion or incision of an arterie it has the same pulsation with an arterie, and yet it has not the tunicle of an arterie. Most learned *Riolan* doth witness this with me in his seventh Book. Nor let any man believe, that the use of pulse and respiration is one and the same, because that the pulses are greater, more frequent, and swifter, for the same causes as respiration is, to wit with running, anger, bathing, or any other thing which heats. For not only that experiment is false (which *Galen* endeavours to convince) that by immoderate repletion the pulses are greater, and breathing lesser; but likewise in boys, pulses are frequent, and respiration the while very seldom. Likewise in fear, care, and anxiety of the mind, as also too in some feavers the pulses are swift and frequent, and respirations more seldome. These and the like inconveniences do follow upon the opinions which are set down concerning the pulse and use of the arteries. Likewise those things which are affirmed concerning the pulse and use of the heart are no less entangled with very many and inextricable difficulties. They do commonly affirm that the heart is the store-house and fountain of vital Spirit, by which it gives life to all the parts, and yet they deny that the right ventricle makes Spirits, but only gives nourishment to the lungs; from whence, say they, fishes have no right ventricle of the heart, and indeed in those which have no lungs it is wanting, and that the right ventricle of the heart was meerly made for the lungs' sake.

1. Why, I beseech you, since the constitution of both the ventricles is alike, their fibers fram'd alike, and so of their tendons, Portals, vessels, ears, and both of them are found full of blood in dissection, alike blackish, alike knotty: why, I say, should we think that they were appointed to such diverse different uses, seeing action, motion, pulse, is the same in both? If the three-pointed portals in the entry of the right ventricle, be a hinderance of the return of the blood into the *vena cava*, and if those three semilunary portals in the orifice of the *arteriosa vena* were made to hinder the regress of the blood; since they are so likewise in the left ventricle, shall we deny that they were likewise made to hinder the egress and regress of the blood there?

2. And since they are almost altogether after the same manner, both in their form and position in the left as in the right, why do they say that here they hinder the egress and regress of the Spirits, and in

9

the right hinder the egress and regress of the blood? This same organ does not seem to be fit to hinder the motion of the blood and Spirits alike.

3. And since the passages and vessels answer to one another in point of size, namely the *vena arteriosa* and *arteria venosa*, why should the one be destined to a particular use, that is to say to nourish the lungs, the other to a general?

4. And how is it probable, as *Realdus Columbus* does observe, that there needs so much blood to the nutrition of the lungs, since this vessel, that is to say the *vena arteriosa*, is bigger than both the branches of the distributives descending into the crural vein?

5. And I beseech you since the lungs are so near, and the vessel is so great, and they in continual motion, what needs the motion of the right ventricle, and what is the matter that nature for the nourishing of the lungs was forc'd to joyn another ventricle to the heart?

When they say that the left ventricle draws matter out of the lungs, and the right bosome of the heart, to make Spirits, that is to say air and blood, and does likewise distribute the spirituous blood into the aorta, and that fumes are sent back by the Venal arterie into the lungs, and the Spirits into the aorta, what is it that makes the separation, or how comes it to pass that spirits and fumes pass sometimes hither sometimes thither without permission and confusion? if the three-pointed mitre-fashioned portals hinder not the return of fumes into the lungs, how shall they hinder the return of air? And how shall the half-moon portals hinder the regress of the spirits from the aorta, the Diastole of the heart pursuing? and by what manner of way do they say that the spirituous blood is distributed through the Venal arterie into the lungs out of the left ventricle, and that the three-pointed doors do not hinder? seeing they affirm that the air does enter through the same vessel out of the lungs into the left ventricle, to the regress of which they would have these three-pointed doors to be a hinderance. Good God, how shall the three-pointed doors hinder the regress of air and not of blood? Further they having destined the *vena arteriosa*, being a large vessel, made with the tunicle of an arterie, for one only and a private use, that is to say to nourish the lungs, Why do they affirm that the Venal arterie being scarce so big, having the tunicle of a vein soft and loose, to be made for more uses, to wit three or four? For they will have the air pass through it out of the lungs into the left ventricle, and they will have the fumes likewise to return through it out of the heart into the lungs, they will have a part of the spirituous blood to be distributed by it, for the refreshing of them: They will have these to send fumes from the heart, and the other to send air to the heart by the same pipe, when notwithstanding nature did not use to frame one vessel and one way for such contrary motions and uses, nor is it ever seen to be so.

If they do affirm that fumes and air do go and

return by this way, as through the transpirations or *Bronchia* of the liver, why cutting up the *arteria venosa* can we find neither air nor fumes? And whence is it that we see that *arteria venosa* alwayes full of thick blood, and never full of air, since we see air remaining in the lungs?

If any would try the Experiment of *Galen*, and cut the windpipe of a dog being yet alive, and forcibly fill the lungs with air, and being filled bind them streight, afterwards cutting up his breast he shall find great store of air in the lungs, even to their utmost tunicle, but nothing in the *arteria venosa*, nor in the left ventricle of the heart. But if in a living dog either the heart did attract it, or the lungs did pulse it through, they should do it much more in this experiment. Yea in the administration of Anatomy blowing up the lungs of a dead body, who doubts but the air would enter this way, if there were any passage? But they do so much esteem the use of this *arteria venosa* for the conveying of air from the lungs to the heart, That *Hieronimus Fabricius ab Aqua pendente* does assert, that the lungs were made for this vessel's sake, and that it is the chiefest part of the lungs.

But I beseech you, if the *arteria venosa* had been made for the conveying of air, why has it the constitution of a vein?

Nature would stand more in need of pipes, and of annular ones, indeed such as the *Bronchia* are, that should be alwayes open and never lie flat, that they might be altogether void of blood, lest the wetness should hinder the passage of the air, as it is manifest (when the Lungs are diseas'd by the stuffing or least entry of flegm into the *Bronchia*) when we make a whistling or a noise in our breathing.

That opinion is less tolerable, which (supposing that an airy and bloody matter is necessary for the making of vital Spirits) does assert, that the blood is drawn through the hidden pores of the mediastin of the heart, out of the right ventricle into the left, and that the air is drawn through a great vessel, the *arteria venosa*, out of the Lungs; and for that cause, that there are more pores in the septum of the heart, fitter for the production of the blood. But by my troth there are no such pores, nor can they be demonstrated.

For the substance of the septum of the heart is thicker, and more compact than any part of the body, except the bones and nerves. But if there were holes, how were it possible, (since both the ventricles are distended at one time) that the one can draw any thing from the other, or that the left can draw blood from the right? And why should not I rather believe that the right draws Spirits from the left, than that the left through the same holes should draw blood from the right? But it is truly wonderfull and incoherent, that at the same instant the blood should be most conveniently drawn through hidden and obscure passages, and air through very open ones. And why, I beseech you, have they their refuge to hidden, invisible, incertain, and obscure pores for the passage of the blood into the left ventricle, when there is such an open way through the *arteria venosa?* Truly it is a wonder to me, that they would rather invent or make a way through the septum of the heart, which is gross, thick, hard, and most compact, than through the patent *Vas Venosum*, or else through the substance of the lungs, thin, loose, most soft and spongious. Besides, if the blood could pass through the substance of the septum, or be imbib'd by the ventricles, what need were there of the branches of the Coronal arterie divided for that purpose? Which is very worthy to be observ'd, if in a Birth (when all things are thinner and softer) Nature was forced to bring the blood through an oval hole, out of the *Vena Cava* through the *Arteria Venosa*, how can it be possible that she should pass it so conveniently, and with no trouble, through the septum of the heart, being now made thicker after growth? *Andreas Laurentius* in his *Lib.* 9, *Chap.* 11, *Quæst.* 12, being back'd with the authority of *Galen, De Locis Affectis, lib. vi, cap.* 7, and the experience of *Hollerius*, affirms, that whey and matter out of the cavity of the brest, being supp'd up by the *Arteria Venosa*, can be expelled through the left ventricle of the heart and the arteries, together with the Urine and the Excrements; As likewise for the confirmation of it he relates the case of a certain Melancholy man who was freed from a Paroxism by the emission of troubled, stinking, tart urine, by which kind of disease at last dying, and dissecting the body, no such substance as he piss'd, did either appear in the bladder or in the reins, any where, but a great deal in the left ventricle of the heart and concavity of the breast, whence he vaunts that he foretold the cause of such diseases. But I cannot chuse but wonder, since he had guess'd and foretold that Heterogeneous matter could be evacuated by the same passage, that he either could not or would not see or affirm, that through the same wayes the blood could be conveniently, according to Nature, brought out of the lungs into the left ventricle.

Therefore from these, and many such things as these, it is clear, that those things which are before spoken by former Authors concerning the motion and use of the heart and the arteries do either seem inconvenient or obscure, or admit of no compossibility, if one do diligently consider them; therefore it will be profitable to search more deeply into the business, and to contemplate the motions of the arteries and heart, not only in man, but also in all other creatures that have a heart; as likewise by the frequent dissection of living things, and by much ocular testimony to discern and search the truth.

ANATOMICAL
EXERCISES CONCERNING
THE MOTION OF THE HEART AND BLOOD
IN LIVING CREATURES.

CHAPTER I.
The Causes which mov'd the Author to write.

WHen first I applyed my mind to observation from the many dissections of Living Creatures as they came to hand, that by that means I might find out the use of the motion of the Heart and things conducible in Creatures; I straightwayes found it a thing hard to be attained, and full of difficulty, so with *Fracastorius* I did almost believe, that the motion of the Heart was known to God alone: For neither could I rightly distinguish, which way the Diastole and Systole came to be, nor when nor where the dilatation and constriction had its existence. And that by reason of the quickness of the motion, which in some creatures appeared in the twinckling of an eye, like the passing of Lightning; so that sometimes the Systole did present it
Contraction. self to me from this place, and the Diastole from that
Extension. place, sometimes just contrary, sometimes the motion was various, sometimes confus'd; whence I was much troubled in mind, nor did I know what to resolve

16

upon my self, or what belief to give to others; nor wonder'd I at that which *Andreas Laurentius* writes, That the motion of the heart was as the ebbing and flowing of *Euripus* to *Aristotle.* At last using daily more search and diligence, by often looking into many and several sorts of creatures, I did believe I had hit the nail on the head, unwinded and freed my self from this Labyrinth, and thought I had gain'd both the motion and use of the heart, together with that of the arteries, which I did so much desire: Since which time I have not been afraid, both privately to my friends, and publickly in my Anatomy Lectures to deliver my opinion.

Which, as it commonly falls out, pleased some, and displeased others: Some there were that did check me, spoke harshly, and found fault that I had departed from the precepts and belief of all Anatomists; Others avouching that it was a thing new, worthy of their knowledge, and exceeding profitable, requir'd it to be more plainly delivered to them. At last, mov'd partly by the requests of my friends, that all men might be partakers of my endeavours, and partly by the malice of some, who being displeas'd with what I said, and not understanding it aright, endeavoured to traduce me publickly, I was forced to recommend these things to the Press, that every man might of me, and of the thing it self, deliver his judgment freely. But so much the more willing I was to it, because *Hieronymus ab Aquapendente* having learnedly and accurately set down in a particular Treatise almost all the parts of living creatures, left the heart only untouched. Lastly, if any profit or advantage might by my industry in this accrew to the republick of Literature, it might perchance be granted that I had done well, and others might believe that I had not spent my time altogether to no purpose, and as the old man says in the *Comedy*,

> *No man so well e'r laid his count to live,*
> *But that things, age, and use some new thing give,*
> *That what you thought you knew, you shall not know,*
> *And what you once thought best, you shall forgo.*

This may perchance fall out now in the motion of the heart, that from hence the way being thus pervious, others trusting to more pregnant wits, may take occasion to do better and search further.

CHAPTER II.

What manner of motion the Heart has in the dissection of living Creatures.

FIRST then in the hearts of all creatures, being dissected whilst they are yet alive, opening the breast, and cutting up the capsule, which immediately environeth the heart, you may observe that the heart moves sometimes, sometimes rests: and that there is a time when it moves, and when it moves not.

This is more evident in the hearts of colder creatures, as the Toads, Serpents, Frogs, House-Snails, Shrimps, Crevisses, and all manner of little Fishes. For it shews it self more manifestly in the hearts of hotter bodies, as of Dogs, Swine, if you observe attentively till the heart begin to die and move faintly, and life is as it were departing from it. Then you may clearly and plainly see that the motions of it are more slow and seldom, and the restings of it of a longer continuance: and you may observe and distinguish more easily, what manner of motion it is and which wayes it is made in the resting of it, as likewise in death, the heart is yielding, flagging, weak, and lyes as it were drooping.

At the motion, and whilst it is moving, three things are chiefly to be observed.

1. That the heart is erected, and that it raises it self upwards into a point, insomuch that it beats the

breast at that time, so as the pulsation is felt outwardly.

2. That there is a contraction of it every way, especially of the sides of it, so that it appears lesser, longer and contracted. The heart of an Eel, taken out and laid upon a trencher, or upon one's hand doth evidence this: It appears likewise in the hearts of little Fishes, and of those colder Animals whose hearts are sharp at top and long.

3. That the heart being grasp'd in one's hand whilst it is in motion, feels harder. This hardness arises from tention, like as if one taking hold of the tendons of one's arm by the Elbow whilst they are moving the fingers, shall feel them bent and more resisting.

4. 'Tis moreover to be observed in Fish and colder Animals which have blood, as Serpents, Frogs, at that time when the heart moves it becomes whitish, when it leaveth motion it appears full of sanguine colour. From hence it seemed to me, that the motion of the heart was a kind of tention in every part of it, according to the drawing and constriction of the Fibers every way; because it appear'd that in all its motions, it was erected, received vigour, grew lesser and harder, and that the motion of it was like that of the muscles, where the contraction is made according to the drawing of the nervous parts and fibers; for the muscles, whilst they are in motion and in action, are envigorated and stretched, from soft become hard, they are uplifted and thickned, so likewise the heart.

CHAPTER II

From which observations with good reason we may gather that the heart at that time whilst it is in motion, suffers constriction, and is thickned in its outside, and so streightned in its ventricles, thrusting forth the blood contained within it: which from the fourth observation is evident, because that in the tention it becomes white, having thrust out the blood contained within it, and presently after in its relaxation and rest a purple and crimson colour returns to the heart. But of this no man needs to make any further scruple, since upon the inflicting of a wound into the cavity of the ventricle, upon every motion and pulsation of the heart, in the very tention, you shall see the blood within contained to leap out.

So then these things happen at one and the same time, the tention of the heart, the erection of the point, the beating (which is felt outwardly by reason of its hitting against the breast), the incrassation of the sides of it, and the forcible protrusion of the blood by constriction of the ventricles.

Hence the contrary of the commonly received opinion appears, which is, that the heart at that time when it beats against the breast and the pulsation is outwardly felt, it is believ'd that the ventricles of the heart are dilated, and replete with blood, though you shall understand that it is otherwise, and that when the heart is contracted it is emptied. For that motion which is commonly thought the Diastole of the heart, is really the Systole, and so the proper motion of the heart is not a Diastole but a Systole, for the heart receives no vigour in the Diastole, but in the Systole, for then it is extended, moveth, and receiveth vigour.

Neither is that to be allowed (though it is confirmed by a comparison alleged by the Divine *Vesalius* of a wreath of Oziers, meaning of many twigs joyn'd together in fashion of a Pyramide) that the heart doth not only move by the streight fibers, and so whilst the top is brought near to the bottom, the sides of it are dilated round about and do acquire the form of a little gourd, and so take in the blood (for according to all the drawing of the Fibers which it has, the heart is stiffned and gather'd together); But that the outside and substance of it are rather thickned and dilated, and that whilst the Fibers are stretched from the top of the corner to the bottom, the sides of the heart do not encline to an orbicular figure, but rather contrary, as every Fiber circularly plac'd does in its contraction encline to streightness, and as all the fibers of the muscles whilst they are contracted and shortned of their length, so towards the sides they are extended, and are thickned after the same fashion as the bodies of the Muscles.

To this add, that not only in the motion of the heart, by erection and incrassation of the sides of it, it so falls out, that the ventricles are streightned, but moreover all the sides inwardly are girt together as it were with a noose for expelling the blood with greater force, by reason that those Fibers or little tendons, amongst which there are none but streight ones, (for those in the outside are circular) called by *Aristotle* Nerves, are various in the ventricles of the hearts of greater creatures, whilst they are contracted together with a most admirable frame.

Neither is it true which is commonly believ'd, that the heart by any motion or distention of its own doth draw blood into the ventricles, but that whilst it is moved and bended, the blood is thrust forth, and when it is relax'd and falls, the blood is received in manner as follows.

CHAPTER III.

What manner of motion the Arteries have in dissection of living creatures.

THERE occurs in the motion of the heart these things further to be observ'd, which have relation to the moving and pulsation of the arteries.

1. That whilst there is a tention, contraction of the heart, and a percussion of the breast, and an apparent Systole, the arteries are dilated, do beat, and are in their Diastole. In like manner when the right ventricle thrusts out the blood contained in it, the arterious vein beats and is dilated, together with the rest of the arteries of the body.

2. When the left ventricle ceaseth to move, beat and to be contracted, the beating of the arteries ceases: nay, when the tention is but faint, the pulsation of the arteries is hardly to be perceived, and so likewise in the arterial vein, when the right ceases.

3. Likewise cutting or piercing any arterie in the very tention of the left ventricle the blood is forcibly thrust out of the wound; so, cutting the arterial vein at the same time and in the tention and contraction of the right ventricle, you shall see the blood to burst out forcibly from thence.

So likewise in Fishes, cutting the conduit pipe, which leads from the heart to the gills, at which time you shall see the heart stiff and contracted, from thence you shall see the blood forcibly thrust out.

Lastly, as in the cutting of any arterie, the blood leaps out sometimes farther, sometimes nearer, you shall find the out-leaping to be just with the Arterial Diastole, at which time the heart strikes the breast, and at that time then when it appears that the heart is in its tention and contraction, it is in its Systole, and that the blood is thrust out with the same motion.

From hence, this against the Common rule appears to be clear, that the Arterial Diastole is at the same time with the Systole of the heart, and that the arteries are fill'd and distended by reason of the immission and intrusion of blood made by the constriction of the ventricles of the heart; as likewise that the arteries are stretched, because they are fill'd like Baggs or Satchels, and are not fill'd because they are blown up like Bellows: and for the same cause do all the arteries of the body beat, by reason of the tention of the left ventricle of the heart, as the arterial vein from the tention of the right.

Lastly, That the pulsation of the arteries arises from the impulsion of blood from the left ventricle; just so, as when one blows into a glove, he shall see all the fingers swell up together and assimulate this pulsation. As also according to the tention of the heart, the pulsations are greater, more vehement, more frequent, swifter, keeping the number, quantity, and order of the beating of the heart.

Nor is it to be expected, that because of the

motion of the blood there should be a certain distance of time betwixt the constriction of the heart and the dilatation of the arteries (especially of those that are furthest distant), or that they be not at the same instant; because that in a Bason (as likewise in a Drum and long pieces of Timber) the stroke and the motion are alike soon at both extremes: since the case here is just as in the blowing up of a Glove, or a Bladder. Hence *Aristotle, De Animalium iii, cap.* 9, *De Respiratione, cap.* 15. *The blood* (says he) *of all living creatures, beats within their veins* (meaning the arteries), *and with a continual motion moves everywhere: so do all the veins beat together, and by turns, because they have their dependance upon the heart. But it does alwayes move, wherefore they likewise move, and in order to its motion when it doth move.*

We must observe with *Galen*, that the arteries were named veins by the antient Philosophers. I chanced on a time to see and have in hand an accident which did most plainly confirm this to me to be true: A certain person had a great swelling which did beat on the right side of his throat near to the descent of the subclavial arterie into the armpits, call'd *Aneurisma*, begotten by the corrosion of the arterie it self, which grew bigger and bigger every day, being filled with the immission of blood from the arterie at every pulsation; which was found upon the cutting up of his body after he was dead. In this man the pulse of his arm upon that side was very weak, by reason that the greater portion and influx of blood was turned into the swelling, and so diverted.

Wherefore, whether it be by compression, stuffing, or interception that the motion of the blood through the arteries be hindered, in that case the furthermost arteries do beat less, seeing the pulse of the arteries is nothing but the impulsion of the blood into the arteries.

CHAPTER IV.

What manner of motion the Heart and the ears of it have in living creatures.

BESIDES these, there are to be observed such things as belong to the ears, which *Gaspar Bauhinus, P. C. Anat., lib.* 2, *cap.* 21, and *Johan Riolanus, Antrop., lib.* 3, *cap.* 12, men very learned, and skilfull Anatomists have observed, and advise us, that if in the live dissection of any animals you have good regard to the motion of the heart, you shall see four motions, distinct both in time and place: with leave of such eminent men be it spoken, there are four motions distinct in place, but not in time; for both the ears move together, and both the ventricles move together, so that there are four motions distinct in place, only at two times, and it is thus.

There are as it were at one time two motions, one of the ears, and another of the ventricles themselves, for they are not just at one instant, but the motion of the ears goes before, and the motion of the heart follows; and the motion seems to begin at the ears, and to pass forward to the ventricles; when all things are already in a languishing condition, (the heart dying away, as it is both in Fishes and other colder animals which have blood) there intercedes some short resting time betwixt these two motions, and the heart being as it were weakned, seems to answer the motion, sometimes swifter, sometimes

slower; last of all drawing towards death, it ceases
to answer by its motion, and only by nodding its
head seems as it were to give consent, and moves
so insensibly, that it seems only to give a sign of
motion to the ears: So the heart first leaves beating
before the ears, so that the ears are said to out-live
it: the left ventricle leaves beating first of all, then
its ear, then the right ventricle, last of all (which
Galen observes) all the rest giving off and dying,
the right ear beats still: so that life seems to remain
last of all in the right. And whilst by little and little
the heart is dying, you may see after two or three
beatings of the ear, the heart will, being as it were
rowsed, answer, and very slowly and hardly endeavour and frame a motion.

But this is chiefly to be observed, that after the
heart has left beating, and the ears are beating still,
putting your finger upon the ventricle of the heart,
every pulsation is perceived in the ventricles, just
after the same manner as we said the pulsations of
the ventricles were felt in the Arteries, a distention
being made by impulsion of blood: and at this time,
the ears only beating, if you cut away the point of
the heart with a pair of Scissors, you shall see the
blood flow from thence at every pulsation of the
ear, so that from thence it appears which way the
blood comes into the ventricles, not by attraction
or distention of the heart, but sent in by the impulsion of the ears.

It is to be observed, that all those which I call

pulsations, both in the ears and in the heart, are contractions, and that the ears are evidently first contracted, and afterwards the heart it self. For the ears whilst they move and beat, become whitish, especially when there is little blood in them, for they are fill'd as the cellars and treasuries of blood, by the compressive motion of the veins, and the tending of the blood to its proper Centre. Nay further, it is most evident, in the ends and extremities of them, that the whiteness arises meerly from the contraction of them.

In Fishes, and Froggs, and the like, having but one ventricle of the heart (for in lieu of one ear they have a little bladder plac'd at the bottom of their heart full of blood) you shall most evidently see the bladder first contracted, and the contraction of the heart to ensue.

Notwithstanding I thought fit to insert those things which were of a contrary course: the heart of an Eel, as also of some Fishes and living creatures, being tane out beats without ears, nay, though you cut it in pieces, you shall see the pieces when they are asunder contract and dilate themselves, so that in such, after the motion of the ears, the heart does leap and beat: But this perchance is only proper to such creatures, which are more tenacious of life, whose radical moisture is more glutinous, fatter, tougher, and not so easie to be dissolv'd. This also does appear in the flesh of Eels, which after the skinning, exenteration, and cutting in pieces, retains motion.

CHAPTER IV

This is certain, that upon a time trying an experiment upon a Dove, after that the heart had quite left motion, and that the ears had a while given over, I wetted my finger with spittle, and being warmed kept it a while upon the heart; by this fomentation, as if it had received strength and life afresh, the heart and its ears began to move, to contract, and open, and did seem as it were recall'd back again from death.

But besides all these I have often observ'd that after the heart it self, and even its right ear, had at the very point of death left off beating, there manifestly remain'd in the very blood which is in the right ear, an obscure motion, and a kind of inundation and beating, that is to say, so long as it seem'd to be possess'd with any blood or spirit.

A thing of the like nature, in the first generation of a living creature most evidently appears in a Hen's egg within seven dayes after her sitting; first of all there is in it a drop of blood, which moves, as *Aristotle* likewise observ'd, which receiving encrease, and the Chicken being form'd in part, the ears of the heart are fashioned, which beating there is always life; then afterwards within a few days the body beginning to receive its lineaments, then likewise is the body of the heart framed, but for some days it appears whitish and without blood, nor doth it beat and move as the rest of the body; as also I have seen in a child after three moneths, the heart to be also form'd, but whitish, and without blood;

31

in the ears of which notwithstanding there was great store of blood, and of a crimson colour: so likewise in the egg when the Chick was new form'd, and encreased, the heart began likewise to encrease, and to have ventricles in which it began to receive blood and pass it through.

So that if a man will more narrowly pry into the truth, he will not say, that the heart is the first thing that lives, and last that dies, but rather the ears (and in Snakes, Fishes and such like creatures, the part which is instead thereof), and that it both lives before the heart, and dies after it.

Nay, it's doubtfull too, whether or not before them also the spirit and blood have an obscure beating, which to me it seem'd to retain after death, or whether we may say that with this beating the life begins, seeing the Sperm and prolifique Spirit of all living creatures goes from them with a kind of leaping, as if it self were a living creature. So Nature in death making as it were a recapitulation, returns upon her self (as *Aristotle* says, *De Motu Animal.*, *cap.* 8) with a retrograde motion from the end of her race to the beginning of it; from whence she first issues thither she returns, seeing the generation of living creatures, from not being a living creature, is to be a living creature, as from a non-entity to be an entity, so by the same steps corruption passes from an entity to a non-entity; whence it is, that that which in living creatures is last made, fails first, and that which is first made, fails last.

I have likewise observ'd, that there is really a heart in all animals, and not only (as *Aristotle* says) in the greater sort, and such as have blood, but likewise in lesser, and such as have none, as those that are crusted without, or have shels, as house-Snails, Crabfish, Crevisses, Shrimps, and in many others; nay, in Wasps, Hornets, and in Gnats, by an optick glass made for the discovery of the least things, in the upper end of that place which is called their tail, I saw the heart beat, and shewed it to others.

But in those creatures which have no blood, the heart beats very slowly, and with deliberate stroaks, as it does in other creatures which are dying, and is contracted leisurely, as in Snails is easie to discern, whose heart you shall find in the right side at the bottom of that Orifice, which it seems to open and shut for taking of air, and from whence it casts out foam, dissecting it at the top near the place which is answerable to the liver.

But it is to be observed likewise, that in Winter and colder seasons, some creatures which have no blood, such as is the Snail, have nothing which beats, but do rather seem to be like plants; as likewise the rest, which for that cause are called Plant-animals. It is likewise to be observed, that in all creatures which have hearts, there are ears likewise, or some thing answerable to them, and wheresoever the heart has two ventricles, there are two ears, but not contrarily. But if you observe the fashioning of a Chick in the egg, first of all there is in it as

I said only a bladder or drop of blood, which beats, and encreasing afterwards the heart is perfected: so in some creatures (as not reaching a further perfection) there is a certain little bladder only like a point, red or white, as the beginning of life, as in Bees, Wasps, Snails, Shrimps, Crevisses.

There is found here with us a sort of very little Fish, called in English a Shrimp, and in Low Dutch *een Garneel,* usually taken in the Sea and in the River of Thames, all the body of which is transparent: This little Fish I have often shewn in water to some of my special friends, so that we could clearly discern the motion of the heart in that creature, the outward parts nothing at all obstructing our sight, as if it had been through a window. In a Hen's egg I shewed the first beginning of the Chick, like a little cloud, by putting an egg off which the shell was taken, into water warm and clear, in the midst of which cloud there was a point of blood which did beat, so little, that when it was contracted it disappeared, and vanish'd out of our sight, and in its dilatation, shew'd it self again red and small, as the point of a needle; insomuch as betwixt being seen and not being seen, as it were betwixt being and not being, it did represent a beating, and the beginning of life.

CHAPTER V

The action and office of the motion of the Heart.

I confidently believe then, that out of these and the like observations, it will be found that the motion of the heart is after this manner.

First of all the ear contracts it self, and in that contraction throws the blood with which it abounds, as the head-spring of the veins, and the cellar and cistern of blood, into the ventricle of the heart, which being full, straightway the heart raises it self, stretches all the nerves, contracts the ventricles, and makes a pulsation: by which pulsation it continually thrusts that blood (which by the ears is sent in) forth into the arteries, the right ventricle into the lungs, through that vessel which is called the *vena arteriosa*, but is indeed both in its place and function and every thing else an arterie; the left ventricle into the aorta, and so by the arteries into the whole body.

Those two motions, the one of the ears, the other of the ventricles, are so done in a continued motion, as it were keeping a certain harmony and number, that they are both done at the same time, and one only motion appears, especially in hotter creatures, whilst they move with a sudden motion. Nor is this otherwise done, than when in Engines, one wheel moving another, they seem all to move together; and in the lock of a piece, by the drawing of the spring, the flint falls, strikes the steel, fires the powder, enters the touch-hole, discharges, the balls fly out, pierces the mark, and all these motions, by reason of the swiftness of them, appear in the twinkling of an eye: So likewise in the deglutition, the meat or drink is thrown into the jaws, the larinx is shut close by its own muscles, and the Epiglottis, the top of the weason, is lifted up and opened by its muscles, just as a sack is raised to be filled, and opened that it may receive; it thrusts down the meat or drink being receiv'd, by the thwarting muscles, and with the long muscles sucks it down; yet notwithstanding that all these motions are made by several and contra-distinct organs, whilst they are done in harmony and order, seem but to make one motion and action, which they call swallowing.

So it comes to pass clearly in the motion and action of the heart, which is a kind of swallowing, and transfusion of blood out of the veins into the arteries. And if any man carefully observing this, shall diligently search the motion of the heart in the dissection of any living thing, he shall see not only that which I have said, that the heart erects it self and makes one continued motion with the ears of it, but likewise a certain motion and inclination side-wayes, and an obscure leaning that way, in order to the draught of the right ventricle, so carrying on the work. As we may see when a Horse drinks and swallows the water, at every gulp the water is sup'd down into the belly, which yields a certain noise and pulse to him that heeds him and touches him; even so it comes to pass, that whilst some portion of the blood is drawn out of the veins into the arteries, there is a beating which is heard within the breast.

The motion of the heart then is after this manner, and the transfusion and propulsion by mediation of the arteries is one of the actions of the heart, so that the pulsation which we feel, is nothing else but only the impulsion of the blood by the heart.

But whether or no the heart contribute any thing else to the blood, besides the transposition, local motion, and distribution of it, we must enquire afterwards, and collect out of other observations. Let this suffice for the present, that it is sufficiently evidenced, that in the beating of the heart the blood is transfused and drawn out of the veins into the arteries through the ventricles of the heart, and so distributed into the whole body.

But this all do in some manner grant and gather from the fabrick of the heart, and from the figure, place and use of the Portals, yet stumbling as it were in a dark place, they seem to be dim-sighted, and clamper up divers things, which are contrary and inconsistent, and speak many things at random (as we shewed before.) One thing seems to me to have been the chief cause of doubt and mistake in this business, which is, the contexture in a man of the heart and lungs; For when they did see the *vena arteriosa*, and the *arteria venosa*, coming likewise into the lungs, and there to disappear, it could not sink with them either how the right ventricle should distribute the blood into the body, or how the left ventricle should draw it out of the *vena Cava*. This *Galen's* words do testifie in his book *De Placitis Hippocratis & Platonis*, 6, where he inveighs against *Erosistratus* concerning the beginning and use of the veins, and the concoction of the blood. *You will answer* (sayes he) *that it is so ordained, that the blood be prepared in the Liver, and so carried to the Heart, there to receive its proper form and absolute perfection: which truly seems not without reason; for no perfect and great work is done suddenly, at one attempt, and gains all its refining from one instrument. Which if it be so, shew us another vessel which draws out the blood, being absolutely perfected from the heart, and disposes of it as the arteries do of the spirits through the whole body.*

See here an opinion, which carries reason with it, left and rejected by *Galen,* because (besides not perceiving the passage) he could not find a vessel which from the heart should distribute the blood into the whole body.

But if at that time in the defence of that opinion (which is now ours, and in all things else agreeable to reason by *Galen's* own confession) one should with his finger have pointed out the great Arterie dispensing the blood from the Heart into the whole body, what would that Divine man, most ingenious, and most learned, have answered? I wonder whether he would have said that the arteries distribute Spirits and not blood? Certainly he should not by this sufficiently have confuted an *Erosistratus*, who did imagine the Spirits to be contained in the arteries only, but should in the mean time contradict himself, and basely deny that which in one of his own books he stiffly maintains to be true, proves it by many and strong arguments, and by experiments demonstrates it, that blood is naturally contain'd in the arteries, and not Spirits.

But if that Divine man, as he does often in the same place, do grant that all the arteries of the body do arise from the great arterie, and it from the heart, and professing likewise that those three-pointed doors plac'd in the Orifice of the Aorta do hinder the return of the blood into the heart, and that nature had never ordain'd them for the best of our intralls, unless it had been for some special Office, I say, if the father of the Physicians should grant all these things, and in the same very words as he does in his forementioned book, I do not see how he could deny that the great arterie was such a vessel as did carry the blood, after it had received its absolute perfection, out of the heart into the whole body: Or perchance he would still continue to be doubtful, (as all the rest since his time to this very day) because not seeing the contexture of the heart with the lungs he was ignorant of the ways by which the Blood could be carried into the arteries, which doubt does not a little perplex the Anatomists when always in dissections they find the *arteria venosa* and the left ventricle full of thick knotty black blood, so that they are forc'd to affirm that the blood swets through the encloser of the heart from the right ventricle to the left; but this way I have sufficiently refuted already, therefore there must another way be prepared and laid open which being found, there can, I imagine, be no difficulty, which can hinder any body from granting and confessing those things which I propounded before of the pulsation of the heart, and dispensation of the blood by the arteries into the whole body.

CHAPTER VI.

By which ways the blood is carried out of the
vena cava into the arteries, or out of
the right ventricle of the heart
into the left.

SINCE it is probable, that the connexion of the
heart with the lungs has given this occasion of
mistake, they are to be blamed in this, who whilst
they desire to give their verdict, to demonstrate
and understand all parts of living creatures, look but
into man only, and into him being dead too, and
so do no more to the purpose than those who,
seeing the manner of Government in one Commonwealth, frame Politicks, or they who, knowing the
nature of one piece of land, believe that they understand agriculture, or as if from one Particular pro-
position they should go about to frame Universal
arguments.

Nevertheless were they but as well practis'd in
the dissection of creatures, as they are in the Anatomy of men's carcases, this business, which keeps
them all in doubt and perplexity, would in my opinion
seem clear without all difficulty.

First of all in Fishes, having but one ventricle of
the heart (as having no lungs), the thing is clear
enough. For it is certain, that it may be confirmed
before our eyes, that the bladder of blood, which

they have at the bottom of the heart, answerable to the ear of the heart, sends the blood into the heart, and that the heart does afterward, through a pipe or artery, or something answering to an artery, openly transfuse it, both by our own view, and also by cutting the arterie, the blood leaping out upon every pulsation of the heart.

You may likewise see the same afterward easily in all other creatures, in which there is but one ventricle only, or something answerable to it, as in the Toad, Frogg, Serpents, house-Snails, which although they are said in some manner to have lungs, because they have a voice (of the frame of whose lungs I have many observations by me, which are not proper for this place) yet from our own eye-sight it is clear, after the same manner in them that the blood by the pulsation of the heart is brought out of the veins into the arteries, the way of it open, patent, manifest, no occasion of doubt or difficulty at all. For the case is just so with them as it might be with a man, the enclosure of whose heart were pierced through, or taken away, and so both the ventricles become one; I believe no man then would doubt which way the blood should go out of the veins into the arteries.

And seeing there are more creatures which have no lungs than there are which have, and more which have but one ventricle than there are which have two, we may very well aver for the most part, and almost in all, that the blood is transfus'd out of the veins into the arteries through the bosom of the heart by an open passage.

But I conceiv'd with my self that it is plainly seen too in those Embryons which have hearts.

In a birth there are four vessels of the heart, the *vena cava*, the *vena arteriosa*, *arteria venalis*, and the aorta, or *arteria magna*, and are otherwise united than in one come to age, which all Anatomists know well enough.

The first touch and union of the *vena cava* with the *arteria venosa*, which comes to pass before the *vena cava* opens it self into the right ventricle of the heart, or sends out the Coronal vein, a little above its out-going from the liver, displays unto us its orifice side-wayes, that is to say, a hole, wide and large, of an oval figure, made through passageable, from the *vena cava* into that Arterie: Insomuch as through that hole the blood may freely and abundantly pass out of the *vena cava* into the *arteria venosa* and the left ear of the heart, and so to the left ventricle. There is moreover against that place which looks towards the *arteria venosa* a membrane thin and hard, like a cover, which afterwards in those which grow to riper years, covering this hole and growing together every way does quite stop it, and takes away almost all sign of it. This membrane*, I say, is so ordained, that hanging loosely with its own weight, it makes way into the lungs and heart, and is turned up, giving passage to the

* Septum.

blood which flows from the *vena cava*, but hinders it from flowing back into the *cava* again. So that from hence we may imagine in an Embryon, that the blood ought continually to pass through this hole into the *arteria venosa* out of the *vena cava*, and so into the left ear of the heart, and after it is enter'd, that it can never return.

The other union is that of the *vena arteriosa*, (which comes to pass after that that vein coming out of the right ventricle is divided into two branches) and it is as it were a third trunk, or arterial conduit-pipe, divers from the two former, from hence crookedly drawn, and perforate into the *Arteria magna*; so that in the dissection of Embryons, there appears as it were two aortas, or two roots of the great arterie. This conduit likewise in those that come to riper age is attenuated by little and little, and fades away, and at last is quite dried up and lost, like the Umbilical vein. This arterial conduit-pipe hath no membrane to hinder the motion of blood backward, or forward, for there are in the orifice of that *vena arteriosa*, of which this conduit-pipe as I said before is a branch, three doors * of the fashion of a Σ, which appear outwardly and inwardly, and do easily give passage to the blood flowing into the right ventricle by this way, but on the contrary hinder any thing which may flow from the arterie or the lungs into the right ventricle, which they shut very close: So that here we have

* Valvulae.

reason to think, that in an Embryon when the heart contracts it self, the blood must alwayes be carried out of the right ventricle into the *arteria magna* by this way.

In answer to that which is commonly spoken, that these two conjunctions, so great, so open, so wide, were made for the nourishing of the lungs, and that in those who arrive to riper age, when the lungs by reason of their heat and motion require more abundant nutriment, they should be tane away and made up, is an invention improbable and inconsistent. And that is likewise false which they say of the heart of an Embryon, that it is idle and does nothing, moves not at all: whence it comes to pass, that Nature was forc'd for the nourishing of the lungs to make those passages; when by our own eyes it is made plain to us, that both in an egg whereon a Hen hath sate, and in Embryons newly cut out of the womb, the heart doth move as in those of riper age; and likewise, that Nature is pressed with no such necessity: Of which motion not only these my eyes have often been Witnesses, but likewise *Aristotle* himself affirms (*Lib. de Spiritu, cap.* 5.): *The pulse* (says he) *appears at the very beginning in the constitution of the heart, which is found in the dissection of living creatures, and by an egg in the forming of the Chick.* But we also observe, that those passages are open and free, as well in men as also in other creatures, not only to the time of the birth, which the Anatomists have observ'd, but likewise

45

many moneths after, yea in some for many years, if not all their life-time, as in the Goose and very many Birds. Which thing perchance did deceive *Botallus*, so that he affirm'd that he had found a new passage for the blood, out of the *vena cava* into the left ventricle of the heart. And I do confess, that when I my self first found this in a Rat of full growth, that I did imagine some such thing. From which it is understood, that in the unripe births of mankind, and likewise in others, in which these unions are not taken away, this very thing falls out, that the heart by its motion brings forth the blood from the *vena cava* openly and by very patent wayes, by the drawing of both its ventricles. For the right receiving the blood from the ear, thrusts it forth through the *vena arteriosa*, and its branch called *canalis arteriosus*, into the great arterie. Likewise, the left at the same time by the mediation of the motion of the ear, receives that blood, which is brought into the left ear through that oval hole from the *vena cava*, and by its tention and constriction thrusts it through the root of the Aorta into the great arterie likewise. So in Embryons whilst the lungs are idle, and have no action nor motion (as if there were none at all) Nature makes use of both the ventricles of the heart, as of one for transmission of blood. And so the condition of Embryons that have lungs and make no use of them, is like to the condition of those creatures which have none at all.

Therefore in these likewise the truth appears as

clearly, that the heart by its pulsation brings forth and transfuses the blood out of the *vena cava* into the great arterie, and by as open ways as if both the ventricles (as I said before) were made pervious to one another, by taking away the partition betwixt them. Therefore seeing for the most part these ways are open in all creatures at some times, which do serve for transmission of blood through the heart, it now remains that we enquire either why in some creatures, as in men, and those hotter and of riper age, we do hold that not to be performed through the substance of the lungs, which nature did before in an Embryon through those passages (at that time when there was no use of lungs), which she seems to have made of force for want of passage through the lungs. Or why it is better that Nature (for Nature always does that which is best) hath altogether shut up those open ways, of which she before made use in the Embryon and in the birth, and in all other creatures does make use of, nor in the lieu of them hath found out any other passage for the blood, but hinders it altogether after this manner.

So then the business is arriv'd to this, that to those who search for the veins in men (by which the blood passes out of the *vena cava* in the left ventricle and into the *arteria venosa*) it were more worthy their pains, and wiselier done, if from the dissection of living creatures they would search the truth, why in greater and more perfect creatures, and those of riper age, nature would rather have the blood to be squeezed through the streyner of the lungs, than through most patent passages, as in other creatures: and then they would understand that no other way nor passage could be excogitated. Whether this be, because that greater and perfecter creatures are hotter, and when they come to be of age, their heat is apter to be suffocated and to be inflamed, and therefore the blood is streyn'd and sent through the lungs that it may be temper'd by breathing in the air upon it, and freed from overheating and suffocation, or some such other thing. But to determine and give a reason of this is nothing else but a search for what the lungs were made. And thus much concerning them and their use, and all manner of cooling, of the necessity and use of air, and the like, of several and different organs made in animals. For this cause, although by observation I have found out a great many things, yet lest I should seem by straying from my purpose, of the motion of the heart, to go besides my intention, and leave my task to confute the business, and decline it, I shall leave these things fitter to be set forth in a Treatise by themselves; and that I may return to my former purpose, I will go on to prove what remains. And first I prove, that in the more perfect Animals and those come to age, as in Man, the blood may pass from the right ventricle of the heart, by the *vena arteria*, into the lungs, and from thence through the *arteria venosa* into the left ear, and from thence into the left ventricle of the heart, and then that it is so.

48

CHAPTER VII.

That the blood does pass from the right ventricle of the heart, through the ſtreyner of the lungs, into the arteria venosa & left ventricle of the heart.

IT is well enough known that this may be, and that there is nothing which can hinder, if we consider which way the water, passing through the substance of the earth, doth procreate Rivulets and Fountains; or if we do consider how sweat passes through the skin, or how urine flows through the streyner of the reins: It is to be taken notice of in those that make use of the waters of the Spaw, or *de la Madonna,* as they call them in Padua, or other brackish or vitriolated waters; or those who in carrowsing swill themselves with drink, that in an hour or two they pisse all this through their bladder. This great quantity ought to stay a while in concoction, it ought to flow through the liver (as they confess that the juyce of the nourishment we receive doth twice a day), so ought it through the veins, through the streyner of the reins, and through the ureters into the bladder.

Those therefore which I hear denying that blood, yea the whole mass of blood, may pass through the substance of the lungs, as well as the nutritive juyce through the liver, as if it were impossible and no

wayes to be believed — it is to be thought that
those kind of men (I speak with the Poet), where
they like, they easily grant, where they like not, by
no means; here where need is, they are afraid, but
where no need is they are not afraid to aver. The
streyner of the liver, and of the reins too, is much
thicker than that of the lungs, because they are far
thinner woven, and of a spongious substance, if they
be compared to the liver and reins.

In the liver there is no impulsive, no strength
forcing; in the lungs the blood is thrust against them
by the impulsion of the right ventricle of the heart,
by which impulsion there must necessarily follow
a distention of the vessels and porosities of the lungs.
Besides, the lungs in respiration rise and fall (*Galen,
De Usu Partium*), by which motion it follows of
necessity, that the porosities of them and their vessels are open'd and shut, as it falls out in sponges,
and all things of a spongy substance when they are
constricted and dilated again. On the contrary, the
liver is at rest, nor is it seen at any time to be so
constricted and dilated.

Last of all, since through the liver, there is none
but affirms, that the juyce of all things we receive
may pass into the *vena cava*, both in Men, Oxen,
or the greatest creatures, and that for this reason,
because it must pass some way into the veins if
there be any nutrition, and there is no other way,
and for that cause they are forced to affirm this. Why
should they not likewise believe this of the passsage

of the blood through the lungs in men come to age, upon the same arguments? And with Columbus, a most skilfull and learned Anatomist, believe and assert the same from the structure and largeness of the lungs; because that the *Arteria venosa*, and likewise the ventricle, are alwayes full of blood, which must needs come hither out of the veins by no other path but through the lungs; as both he and we from our words before, our own eye-sight, and other Arguments, do believe to be clear.

But seeing there are some such persons which admit of nothing, unless there be an authority alleged for it, let them know, that the very same truth may be proved from *Galen's* own words, that is to say, not only that the blood may be transfused out of the *vena arteriosa*, into the *arteria venosa*, and thence into the left ventricle of the heart, and afterwards transmitted into the arteries, but also that this is done by a continued pulse of the heart, and motion of the lungs, whilst we breath. There are in the orifice of the *vena arteriosa* three shuts, or doors, made like a Σ or half-Moon, which altogether hinder the blood sent into the *vena arteriosa* to return to the heart, which all know.

Galen expresses the use and necessity of those shuts in these words (*De usu partium, lib. 6, cap.* 10.) *In all* (sayes he) *there is a mutual Anastomosis or opening of the veins, together with the arteries, in their kissing, and they borrow both blood and spirit from one another by invisible and very narrow passages. But if*

51

the very mouth of the Vena Arteriosa *had always stood open, and Nature had found no device to shut it when it was requisite, and to open it again, it could never have come to pass that by those invisible and little kisses, the Thorax being contracted, the blood could be transfused into the arteries. For everything is not from any thing extracted and emitted after the same manner; for as that which is light is easilier attracted than that which is heavy, by dilatation of the instruments and by the constriction is squeezed out again; so any thing is easier attracted through a broad passage than through a narrow passage, and so sent forth again. But when the Thorax is contracted, the* Arteriae venosae *which are in the Lungs, being on every side pulsated and compress'd together strongly, do squeeze out very quickly the spirit that is in them, and do borrow through those fine touches a part of the blood, which truly could never come to pass, if through that great opening, such as is the* Vena Arteriosa, *the blood could return back to the Heart: Now the return of it through that great mouth being stop'd, some of it through those small orifices does drop into the Arteries, it being press'd every way.* And a little after in the following Chapter, *How much the more the Thorax endeavours to squeeze out the blood, so much the more those Membranes, that is to say those three Sigmalike doors, do closlier shut the mouth of it, and suffer nothing to return.* Which he says likewise in the same tenth Chapter a little before, *Unless there were doors there would follow a three-fold inconvenience, for so the blood should make such a long journey but in vain, by flowing in the Diastoles of the Lungs, and filling all the veins in them, in the Systoles, as it were a neap tide, like* Euripus *reciprocating its motion again ond again, hither and thither, which would not be convenient for the blood: But this may seem no great matter; but that in the meantime it should weaken the benefit of respiration, this is no more to be counted a small business.* And a little after, *And likewise the third inconvenience would follow, no slight one, when in our breathing our blood should return backwards, unless our* Maker *had ordain'd the natural position of those* Membranes. Whence he concludes, Chap. II, *Indeed the use of all the shuts or portals is the same, to hinder the return of the matter; and either of them have a proper use to draw matter from the heart, that they may return no more, and to draw matters into the heart that they may go no more from thence. For* Nature *would not have the heart to be wearied with needless travel, nor send thither whence it was better to extract, nor extract from thence again whither it was better to send. For which cause, there being four orifices onely, two in either* Ventricle, *one takes in, the other draws forth.* And a little after, *Furthermore, when one of the vessels consisting but of one* Tunicle *is implanted into the* Heart, *and the other consisting of a double* Tunicle *is drawn forth from it, viz.* (The right ventricle *Galen* means, so do I the left ventricle by the same reason) *It was needful that there should be as it were a cistern to both, to which both of them belonging, that the blood might be drawn out by one, and sent out by the other.*

That argument which *Galen* brings for the passages of the blood through the right ventricle out of the *vena cava* into the lungs, we may more rightly use for the passages of the blood out of the veins through the heart into the arteries, changing only the terms.

It does therefore clearly appear from the words and places of *Galen*, a divine man, father of Physicians, both that the blood doth pass from the *vena arteriosa* into the little branches of the *arteria venosa*, both by reason of the pulse of the heart and also because of the motion of the lungs and thorax. (See the commentarie of the most learned *Hofmannus* upon the sixth Book of *Galen*, *De usu partium*, which book I saw after I had written these things.)

Furthermore it was necessary that the heart should receive the blood continually into the ventricles, as in a pond or cistern, and send it forth again : and for this reason it was necessary that it should be serv'd with four locks or doors, whereof two should serve for the intromission and two for the emission of blood, lest either the blood like an *Euripus* should inconveniently be driven up and down, or go back thither from whence it were fitter to be drawn, and flow from that part to which it was needful it should have been sent, and so should be wearied with idle travel and the breathing of the lungs be hindred. Lastly our assertion appears clearly to be true, that the blood does continually and incessantly flow through the porosities of the lungs, out of the right ventricle into the left, out of the *vena cava* into the *arteria magna*; for seeing the blood is continually sent out of the right ventricle into the lungs through the *vena arteriosa*, and likewise is continually attracted out of the lungs into the left, which appears by that which has been spoken, and the position of the Portals, it cannot be, but that it must needs pass through continually.

And likewise seeing that always, and without intermission, the blood enters into the right ventricle of the heart, and goes out (which is likewise manifest of the left ventricle, both by reason and sense), it is impossible but that the blood should pass continually through, out of the *vena cava* into the Aorta.

That therefore which is apparent to be done in most, and really in all whilst they are growing to age, by dissection through most open passages, is here likewise manifest to come to pass in those when they are arriv'd to full age, by the hidden porosities of the lungs, and touches of its vessels both by *Galen's* words, and that which has been spoken: From whence it appears, that albeit one ventricle of the heart, that is the left, were sufficient for the dispensation of the blood through the whole body, and the eduction of it out of the *vena cava* (as it is in all creatures which want lungs), Yet Nature desiring that the blood should be strained through the lungs, was forc'd to add the right ventricle, by whose pulse the blood should be forc'd through the very lungs out of the *vena cava* into the receptacle of the left ventricle: and so it is to be said that the left ventricle was made for the lungs' sake and not for nutrition only; seeing in such an abundance of victual, adding to it the help of compulsion, it is no ways to be believ'd that the lungs should rather want so much aliment, and that of blood so much more pure and full of spirit, as being immediately conveyed from the ventricles of the heart, than either the most pure substance of the brain, or the most resplendent and divine constitution of the eyes, or the flesh of the heart it self, which is more fitly nourished by the *vena coronalis*.

CHAPTER VIII.

Of the abundance of blood passing through the Heart out of the veins into the arteries, and of the circular motion of the blood.

THus much of the transfusion of the blood out of the veins into the arteries, and how it is disposed of and transmitted by the pulse of the heart, to some of which those perchance that were heretofore moved by the reasons of *Galen*, *Columbus*, and others, will yeeld; now as concerning the abundance and increase of this blood, which doth pass through, those things which remain to be spoken of, though they be very considerable, yet when I shall mention them, they are so new and unheard of, that not only I fear mischief which may arrive to me from the envy of some persons, but I likewise doubt that every man almost will be my enemy, so much does custome and doctrine once received and deeply rooted (as if it were another Nature) prevail with every one, and the venerable reverence of antiquity enforces: Howsoever, my resolution is now set down, my hope is in the candor of those which love truth, and learned spirits. Truly when I had often and seriously considered with my self, what great abundance there was, both by the dissection of living things, for experiment's sake, and the opening of arteries, and many ways of searching, and from the Symetrie and magnitude of the ventricles of the heart, and of the vessels which go into it, and go out from it, (since Nature, making nothing in vain, did not allot that greatness proportionably to no purpose, to those vessels) as likewise from the continued and carefull artifice of the doores and fibers, and the rest of the fabrick, and from many other things; and when I had a long time considered with my self how great abundance of blood was passed through, and in how short time that transmission was done, whether or no the juice of the nourishment which we receive could furnish this or no: at last I perceived that the veins should be quite emptied, and the arteries on the other side be burst with too much intrusion of blood, unless the blood did pass back again by some way out of the veins into the arteries, and return into the right ventricle of the heart.

I began to bethink my self if it might not have a circular motion, which afterwards I found true, and that the blood was thrust forth and driven out of the heart by the arteries into the habite of the body and all parts of it, by the beating of the left ventricle of the heart, as it is driven into the Lungs through the *vena arteriosa* by the beating of the right, and that it does return through the little veins into the *vena cava*, and to the right ear of the heart, as likewise out of the lungs through the aforesaid *arteria venosa* to the left ventricle, as we said before.

Which motion we may call circular, after the same manner that *Aristotle* sayes that the rain and the air do imitate the motion of the superiour bodies. For the earth being wet, evaporates by the heat of the Sun, and the vapours being rais'd aloft are condens'd and descend in showrs and wet the ground, and by this means here are generated, likewise, tempests, and the beginnings of meteors, from the circular motion of the Sun and his approach and removal.

So in all likelihood it comes to pass in the body, that all the parts are nourished, cherished, and quickned with blood, which is warm, perfect, vaporous, full of spirit, and, that I may so say, alimentative; in the parts the blood is refrigerated, coagulated, and made as it were barren, from thence it returns to the heart, as to the fountain or dwelling-house of the body, to recover its perfection, and there again by naturall heat, powerfull and vehement, it is melted, and is dispens'd again through the body from thence, being fraught with spirits, as with bals-am, and that all the things do depend upon the motional pulsation of the heart.

So the heart is the beginning of life, the Sun of the Microcosm, as proportionably the Sun deserves to be call'd the heart of the world, by whose vertue and pulsation, the blood is mov'd, perfected, made vegetable, and is defended from corruption and mat-tering; and this familiar houshold-god doth his duty to the whole body, by nourishing, cherishing, and vegetating, being the foundation of life and author of all. But we shall speak more conveniently of these in the speculation of the finall cause of this motion.

Hence it is, seeing the veins are certain ways or vessels carrying the blood, there are two sorts of them, the *Cava* and *Aorta*. Not by reason of the side, as *Aristotle* says, but by their funétion; and not, as is commonly spoken, by their constitution, seeing in many Creatures (as I have said) a vein differs not from an arterie in the thickness of the Tunicle, but by their use and employment distinguishable, a vein and an arterie, both of them not undeservedly called veins by the Antients, as *Galen* has observed, because that this, viz. the arterie, is a way carrying the blood from the heart into the habit of the body, the other a way carrying it from the habit of the body back again into the heart. This is the way from the heart, the other the way to the heart. This contains blood rawish, unprofitable, and now made unfit for nutrition, the other blood digested, perféct, and alimentative.

CHAPTER IX.

That there is a Circulation of the blood from the confirmation of the first supposition.

BUT lest any should think that we put a cheat upon them, and bring only fair assertions without any ground, and innovate without a cause; there comes three things to be confirm'd, which being set down, I think this truth must needs follow and be apparent to all men.

1. First, That the blood is continually, and without any intermission, transmitted out of the *vena cava* into the arteries, in so great abundance, that it cannot be recruited by those things we take in, and insomuch that the whole mass of blood would quickly pass through.

2. In the second place, that continually, duly, and without cease, the blood is driven into every member and part, and enters by the pulse of the arteries, and that in far greater abundance than is necessary for nourishment, or than the whole mass is able to furnish.

3. And likewise thirdly, that the veins themselves do perpetually bring back this blood into the mansion of the heart.

These things being prov'd, I think it will appear that it doth go round, is returned, thrust forward, and comes back from the heart into the extremities, and from thence into the heart again, and so makes as it were a circular motion.

Let us suppose how much blood the left ventricle contains in its dilatation when it's full, either by our thought or experiment, either ℥ij, or ℥iij, or ℥jß.; I have found in a dead man above ℥ij.

Let us suppose likewise, how much less in the contraction, or when it does contract it self, the heart may contain, and how much less capacious the ventricle is, and from thence how much blood is thrust out of the *arteria magna*: for in the Systole there is alwaies some thrust forth, which was demonstrated in the third Chapter, and all men acknowledge, being induced to beleeve it from the fabrick of the vessels, by a very probable conjecture we may aver that there is sent in of this into the arterie a fourth, or fifth, or sixth, at least an eighth, part. So let us imagine, that in a Man there is sent forth in every pulse of the heart, an ounce and a half, or three drams, or one dram of blood, which by reason of the hindrance of the portals cannot return to the heart.

The heart in one half hour makes above a thousand pulses, yea in some, and at some times, two, three or four thousand; now multiply the drams, either a thousand times three drams, or two drams, or five hundred ounces, or such a proportionate quantity of blood, transfus'd through the heart into the arteries, which is a greater quantity than is found in the whole body. So likewise in a Sheep or a Dog if there pass (I grant ye) but one scruple, in one half hour there passes a thousand scruples, or about three pounds and a half of blood: in whose body for the most part is not contained above four pounds of blood, for I have tryed it in a Sheep.

So our account being almost layd, according to which we may guess the quantity of blood which is transmitted, counting the pulsations, it seems that the whole mass of blood does pass out of the veins into the arteries through the heart, and likewise through the lungs.

But grant that it be not done in half an hour, but in a whole hour, or in a day, be it as you will, it is manifest that more blood is continually transmitted through the heart, than either the food which we receive can furnish, or is possible to be contain'd in the veins. Nor is it to be said, that the heart in its contraction sometimes does thrust out, sometimes not, or as much as nothing, or something imaginary. This I refuted before, and besides it's against sense or reason. For as in the dilatation of the heart it must needs come to pass that the ventricles are filled with blood, it is likewise necessary that in its contraction it should alwayes thrust forth, and that not a little, seeing the conduits are not small and the protrusion not seldome: it's very convenient likewise in every propulsion, the proportion of the blood thrust out should be a third part, or sixth part, or eighth part in proportion to that which is before contain'd in the ventricle, and which did fill it in the dilatation, according as the proportion of the ventricle being contracted is to the proportion of it being incontracted; and as in the dilatation it never comes to pass, that it is ever fill'd with nothing, or something meerly imaginary, so in the contraction it never expells nothing, or that which is imaginary, but alwayes something, according to the proportion of the contraction. Wherefore it is to be concluded, that if in a Man, a Cow, or a sheep, the heart doth send forth one dram, and that there be a thousand pulses in one half hour, that it shall come to pass in the same time that there shall be ten pounds and five ounces transmitted, if at one pulse it send forth two drams, twenty pound and ℥ 10, if half an ounce, forty one pounds and ℥ 8, if an ounce, 83 lb. and ℥ 4 will come to be transfus'd, I say, in half an hour, out of the veins into the arteries.

But it may perchance be that I shall set down here more acurately how much is thrust out at every pulsation, when more, and when less, and for what reason, out of many observations which I have gathered.

In the meantime this I know and declare to all men, that sometimes the blood passes in less, sometimes in more abundant quantitie, and the circuit of the blood is perform'd sometimes sooner, sometimes slower, according to the age, temperature, external and internal cause, accidents natural or innatural, sleep, rest, food, exercise, passions of the mind, and the like.

But howsoever, though the blood pass through the heart and lungs in the least quantitie that may be, it is convey'd in far greater abundance into the arteries, and the whole body, than it is possible that it could be supplyed by juice of nourishment which we receive, unless there were a regress made by its circuition.

This likewise appears by our sense, when we look upon the dissection of living things, not only in the apertion of the great arterie, but (as *Galen* affirms in man himself) if any, yea the least arterie be cut, all the mass of blood will be drain'd out of the whole body, as well out of the veins as out of the arteries, in the space of half an hour.

Likewise Butchers can well witness this, when in killing of an oxe they cut the jugular arteries, they drain the whole mass of blood in less than a quarter of an hour, and empty all the vessels, which we find likewise to come to pass in cutting off members and tumours, by too much profusion of blood, sometimes in a little space.

Nor does it weaken the force of this argument, that some will say, that in slaughter, or in cutting off members, the blood flows out as much through the veins as through the arteries, seeing the business is far otherwise. For the veins, because they flap down, and that there is no out-driving force in them, and because their composition is likewise with stoppages of portalls, as hereafter shall appear, they shed but a very little, but the arteries pour out the blood more largely, impetuously, by impulsion, as if it were cast out of a spout. But let the case be tryed omitting the vein and cutting the jugular arterie in a sheep, or a dog, it will be wonderfull to see, with how great force, how great protrusion, how quickly, you shall see all the blood to be emptied from the whole body as well as from the veins as from the arteries. But it is manifest by what we have said, that the arteries receive blood no where else but from the veins by transmission through the heart, wherefore tying the aorta at the root of the heart, and opening the jugular or any other arterie, if you see the arteries empty and the veins only full, it is not to be wondred at.

Hence you shall plainly see the cause in Anatomie why so much blood is found in the veins, and but a little in the arteries, why there is a great deal found in the right ventricle, and but a little in the left (which thing perchance gave occasion of doubt to the antients, and of beleeving that spirits alone were contain'd in those concavities, whilst the animal was alive); the cause perchance is, because there is no passage afforded from the veins into the arteries but through the lungs and the heart, but when the lungs have expir'd and leave off to move, the blood is hindr'd to pass from the little branches of the *vena arteriosa* into the *arteria venosa*, and so into the left ventricle of the heart (as in an Embryon it was before observed, that it was stopt by reason of the want of motion of the lungs, which open and shut up the touches and hidden and invisible porosities), but seeing the heart does not leave off motion at the same time with the lungs, but does beat afterwards and outlive them, it comes to pass that the left ventricle and the arteries do send forth blood into the habit of the body, and not receiving it through the lungs, do therefore appear empty.

But this likewise affords no small credit to our purpose, since there can be no other cause given for this but what in our supposition we have alleged.

Besides, from hence it is manifest, that how much the more, or more vehemently the arteries do beat, it happens in all fluxes of blood that so much the sooner the whole body is emptied.

Hence likewise it comes to pass, that in all faintings, all fear, and the like, when the heart beats more weakly, languishing, and with no force, that it happens that all fluxes of blood are stop'd and hindred.

Hence likewise it is that in a dead body after the heart ceases to beat, you cannot out of the jugular or crurall veins and opening of the arteries by any means extract above half the mass of blood, nor can a butcher when he hath knockt the oxe on the head, and stun'd him, draw all the blood from him unless he cut his throat before the heart leaves beating.

Last of all, from hence we may imagine that no man hitherto has said any thing aright concerning the Anastomosis, where it is, how it is, and for what cause; I am now in that search.

CHAPTER X.

The first supposition concerning the quantitie of the blood which passes through from the veins into the arteries, and that there is a circulation of the blood is vindicated from objections, and further confirm'd by experiments.

THus far the first position is vindicated, whether the matter be to be reckoned by account, or whether we refer it to experiment, or our own eyesight, viz. that the blood continually passes out of the veins into the arteries in greater abundance than can be furnished by our nourishment, so that the whole mass in a little time passing through that way, it must necessarily follow that there should be a circulation, and that the blood should return.

But if any here can say that it can pass through in great abundance, and yet it is not needfull that there should be a circulation, since it comes to be made up by what we receive, and that the encrease of milk in the paps may be an instance, for a cow in one day gives three, four, or seven gallons, or more, a woman likewise gives two or three pints every day or more in the nursing of a child or two, which is manifest to be restor'd by what she receives, it is to be answer'd, that the heart is known to send out so much in one hour or two.

But if not as yet satisfyed he shall still press further, and say, that although by the dissecting of an arterie, and giving and opening a way, it comes to pass besides the course of Nature, that the blood is forcibly pour'd out, yet it does not therefore come to pass in an entire body, no out-let being given, and the arteries being full, and constituted according to Nature, that such a great quantity should pass in so short space, insomuch that there must needs be a regress; It is to be answer'd, That by laying of an account it appears from former reckoning, that how much the heart being fill'd does contain more in its dilatation, than in its constriction, so much (for the most part) at every pulsation is sent forth, and for that cause does there so much pass the body being whole, and constituted according to Nature.

But in Serpents, and in some Fishes, binding the veins a little beneath the heart, you shall quickly see the distance betwixt the heart and the ligature to be emptied, so that you must needs affirm the recourse of blood, unless you will deny your own eye-sight. The same shall clearly appear afterwards in the confirmation of the second supposition.

Let us conclude, confirming all these with one example, that every one may beleeve his own eyes: If any one cut up a live Adder, he shall see the heart beat calmly, distinctly, for a whole hour, and so contract it self (in its constriction being oblong) and thrust it self out again like a Worm. That it is whitish in the Systole, and contrary in the Diastole, together with all the rest, by which I said this truth was evidently confirmed, for here the parts are longer and more distinct. But this we may more especially find, and clearer than the noon-day.

The *vena cava* enters the lower part of the heart, the arterie comes out at the upper part; now taking hold of the *vena cava* with a pair of pinsers, or with your finger and thumb, and the course of the blood being stop'd a little way beneath the heart, you shall upon the pulse perceive to be presently almost emptyed that place which is betwixt your fingers and the heart, the blood being exhausted by the pulse of the heart; and that the heart will be of a far whiter colour, and that it is lesser too in its dilatation for want of blood, and at last beats more faintly, insomuch that it seems in the end as it were to die; so soon again as you untie the vein both colour and bigness returns to the heart. Afterwards, if you do leave the veins, and do grasp or bind the arterie a little way from the heart, you shall on the contrary see them swell vehemently there where they are grasp'd, and that the heart is swell'd beyond measure, and does acquire a purple colour till it be blackish again, and that it is at last opprest with blood so that you would think it would be suffocated, but untying the string, that it does return to its normal constitution, colour, and bigness.

CHAPTER X

So now there are two sorts of death, extinction by reason of defect, and suffocation by too great quantity: here you may have the Example of both before your eyes, and confirm the truth which hath been spoken concerning the heart, by your own view.

CHAPTER XI.

The second supposition is confirmed.

THE second is to be confirm'd by us, which that it may appear the clearer to our view, some experiments are to be taken notice of, by which it is clear, that the blood doth enter into every member through the arteries, and does return by the veins, and that the arteries are the vessels carrying the blood from the heart, and that the veins are the vessels and wayes by which the blood is return'd to the heart it self; and that the blood in the members and extremities does pass from the arteries into the veins (either mediately by an Anastomosis, or immediately through the porosities of the flesh, or both wayes), as before it did in the heart and thorax out of the veins into the arteries: whence it is manifest, that in its circulation it moves from thence hither, and from hence thither, to wit, from the centre to the extremities, and from the extremities again to the centre.

But likewise computation being afterwards made, it appears in the same place, that in regard of the abundance it can neither be recruited by that which we take in, nor is there so much requir'd for nourishment. As likewise concerning ligatures, it is clear how they attract, that they do it not either by heat, nor grief, or force of *vacuum*, nor any other cause known heretofore. As likewise what convenience and use ligatures do bring to Physick, how they stop, or provoke the flux of blood, and how they cause gangrenes and mortifications of the members, and by this means how they are of use in the gelding of some creatures, and in taking away of fleshy tumors and wens. For certainly from hence it comes to pass, that none have rightly understood the causes, and reasons of all these things, though all almost, according to the opinion of the Antients, do propound and give their verdict for ligatures in diseases, yet few in the administration of them do afford any help by them in their cures.

Some ligatures are strict; others of a middle sort.

A strict ligature I call such a one, where the arm is so streightly bound with the band or rope, that you cannot perceive the arterie to beat any where beyond the ligature; such a one we use in the cutting off of members, taking a care of the flux of blood in gelding of animals, taking away of tumors: by which ligature the afflux of aliment and heat being altogether intercepted, we see the testicles to fade and dy, and the great tumors of flesh, and afterwards to fall quite away.

That I call a middle sort of ligature, which does compress the member every way, but without pain, insomuch that it suffers the arterie to beat a little beyond the ligature; such a one as is used in the attraction and emission of blood: for albeit you make the ligature above the elbow, yet you shall perceive the arteries to beat a little in the wrist if you touch it, if in the blood-letting the ligature be made aright.

Now let there be an experiment made in a man's arm, either taking a band, such as they use in blood letting, or by the stronger grasp of the hand it self, which indeed is most conveniently done in a lean body which has larger veins, and when the body being heated, the extremities are warm, and a greater quantitie of blood is in the extremities, and more vehement pulsations, for then all things will more evidently appear.

If you do make then a hard ligature, drawing it as streight as any can endure it, you may first observe that beyond that ligature the arterie does not beat in the wrist, nor any where else, and then that immediately the arterie begins above the ligature, has its Diastole higher, and beats more vehemently, and does as it were with a kind of tide rise towards the ligature (as if it did indeavor to beat through and open its flux which is intercepted) and the passage which is stopt, and that it does appear to be fuller there than is convenient. In the mean time the hand retains its colour and constitution, only in process of time it begins to be a little coldish, but nothing is attracted into it.

After that this ligature has continued a while, and that in a sodain it is a little untied into a middle sort, such I say as they use in letting of blood, it is to be observed that the whole hand is streightways imbued with colour, and distended, and that the veins of it become swell'd and lumpie, and that in the space of ten or twelve pulses the blood being thrust forward and cast into the hand is seen to be extreme full, and that a great quantitie of blood is quickly drawn by the ligature, without either anguish, heat, or shunning of the *vacuum*, or any other cause heretofore mentioned.

In the mean time, if any one put his finger to the arterie, in the very time of the unbinding, near to the ligature, he shall feel the blood as it were passing by under his finger.

Moreover, he in whose arm the experiment is made, upon the change of a streight ligature into a middle one (the impediment being as it were removed) he shall plainly feel the heat and blood enter by pulsation, and perceive something to be breathed by the conduct of the arterie as it were immediately, and to be dispersed over all his hand, and that his hand is presently heated and distended. As in a strict ligature the arteries above are distended, and do beat, and not below, and the veins become lesser, so in the middle sort of ligature the veins swell, and become stubborn, but not above, and the arteries become less, nay if you squeeze the veins, unless you do it very strongly, hardly shall you see the blood pass above the ligature, or the veins fall.

So from these things it is easie for any man that will diligently observe, to know that the blood does enter by the arteries, for by their strict ligature, nothing is attracted, the hand retains its colour, nor

75

happens there any distension, but being a little untied as in the middle or gentle ligature, it is manifest that the hand is swell'd, and that the blood by the force and impulsion is abundantly thrust in. Where the blood flows forth, as in the gentle ligature, they beat; where it does not flow, they beat not at all. In the mean time the veins being streightned nothing can flow through them, of which this is a token, that beneath the ligature they become much more swell'd than above, and than they use to be when the ligature is taken away, hence it is clearly manifest, that the ligature hinders the return of the blood through the veins into the superiour parts, and makes those beneath the ligature continue swell'd.

But the arteries in this case do thrust out the blood beyond the ligatures from the inward parts by the strength and impulsion of the heart, notwithstanding the gentle ligature. This is the difference of the strict ligature from the gentle one, that the strict ligature does not only intercept the passage of the blood in the veins but in the arteries also, that which is gentle doth not hinder the pulsifick vertue, but that it stretches it self and drives out the blood into the furthest parts of the body.

So that we may reason thus: when in a gentle ligature we see the veins swell'd and distended, and the hand to be very full of blood, whence comes this? For either the blood comes through the veins, or through the arteries beneath the ligature, or through the hidden pores. Out of the veins it cannot, by hidden passages less, therefore needs must it by the arteries, as we have said. That it cannot by the veins is apparent, when the blood cannot be squeezed back above the ligature, unless you take the ligature quite away: Then you may see the veins fall and disburthen themselves into the upper parts, and the hand grow white, and all the formerly gathered swelling and blood to vanish apace. He himself will better perceive it, whose body or arm has been so bound a good while, and his hands by that means become swell'd, and made colder; I say, he shall feel somewhat that is cold to creep up to his elbow or armpits, to wit, with the return of the blood, which return of cold blood to the heart after bloodletting, after the untying of the band, I did imagine to be the cause of fainting, which we likewise see come to pass in strong men, and most after the untying of the ligature, which commonly they say comes to pass from the turning of the blood. Besides, when presently upon the untying of the strict ligature into a gentle one, we see, that by the immission of blood through the arteries, the veins comprehended beneath the ligature do swell up, and not the arteries, it is a sign that the blood does pass out of the arteries into the veins, and not on the contrary; and that there is an Anastomosis of the vessels, or that the pores of the flesh and solid parts are pervious to the blood. It is likewise a sign that very many veins do communicate together, when a gentle ligature being made about the arm many of them do swell together, but passage being open'd out of one little vein with the Lancett, they straightwayes fall all of them, and disburthening themselves all into that one, do almost all flap down. From hence may every body know the cause of attraction which is made by ligature, and perchance of all fluxes, viz. as in the hands, when the veins are drawn together by that ligature which I call gentle, the blood cannot go forth; in the mean time if it be driven violently through the arteries, that is to say, by the force of the heart, of necessity the part must be fill'd and distended.

For otherwise how could it be? For heat, anguish, and force of the *vacuum* do indeed attract, but so as the part may be full, not that it should be distended, and swoln beyond its natural constitution. But for the in-thrusting, and straight in-driving of the blood, it is neither to be beleev'd nor can it be demonstrated a member can be suddenly oppress'd, the flesh suffer a solution of its *continuum*, and the vessels be seen to burst, that this can either be done by anguish, heat, or force of the *vacuum*.

Moreover it so falls out, that there is an attraction made by the ligature, without all grief, heat, or force of the *vacuum*. But if by any anguish the blood should chance to be attracted, which way should, beneath the ligature, the hands and the fingers, and the veins swell, and become swell'd, the arm being tyed at the elbow, seeing that by reason of the compression of the ligature the blood could not come thither through the veins? and why should there no sign appear above the ligature either of tumour or repletion, neither any sign of attraction or a flux at all?

But this is the manifest cause of attraction beneath the ligature, and of swelling beyond measure in the hand and fingers, to wit, that the blood does enter forcibly and apace, but cannot get out again.

Hence is all the cause of tumour, and of all oppressive redundancie in any part; because the wayes of ingress are open, and the wayes of regress shut: hence it must needs follow, that the humour should abound, and the part be raised with swelling.

Whether may it not be from hence that in swellings which are inflam'd, so long as the swelling receives increase, and is not in its highest estate, there is a full pulse felt in that place, especially in hotter tumors, in which the increase uses to be on a sudden, shall be for our after-search; as likewise whether that happens from hence, which by chance I had experience of in my self. I falling out of a Coach, and being somewhat hurt in my forehead, there where the little branch of the arterie creeps out of the temples, I felt a swelling about the bigness of an egg in the space of twenty pulses, without either heat or much pain, viz. because of the nearness of the arterie, the blood was abundantly and more swiftly driven into the bruz'd place.

Hence it does appear for what cause in Phlebotomie when we would have the blood leap out further and with greater force, we bind it above the cutting of the vein, not below; but if it flow in so great quantity through the veins from the superiour parts, that ligature would not only not help, but hinder: for it were more likely that it should be bound below, that the blood being hinder'd might go out more abundantly if it did flow thither, and descend from the upper parts into the veins. But since from somewhere else, it is driven by the arteries into the lower veins, in which regress by reason of the ligature is hindred, the veins swell and can squeeze it out, and throw it further through the orifice; but see, the ligature being unty'd, and the way of egress being open, the blood doth no longer come but drop by drop, and that which every body knows, If in Phlebotomy you either untie the band, or bind it below, or bind the member with too strict a ligature, it comes not forth, as if all force were taken from it, because forsooth the way of entrance and influx of blood through the arteries is by that strict ligature intercepted, or a more free regress is granted through the veins, the ligature being untied.

CHAPTER XII.

That there is a circulation of the blood, from the confirmation of the second supposition.

SEEING these things are so, it is certain that another thing which I said before is likewise confirm'd, that the blood does continually pass through the heart. For we see in the habit of the body, that the blood flows continually out of the arteries into the veins, not out of the veins into the arteries: We see, moreover, that from one arm the whole mass of blood may be exhausted, and that too by opening but one cuticular vein with a lance, if the ligature be handsomly made: We see besides, that it is powred out so forcibly and so abundantly, that it is certain that not only that which was comprehended in the arm beneath the ligature, before the section, is quickly and in a little time evacuated, but likewise the blood out of the whole body, as well the veins as the arteries.

Wherefore we must confess first that by strength and force it is furnish'd, and by force it is driven beyond the ligature (for with force it goes out, and therefore by the strength and pulse of the heart) for the force and impulsion of the blood is only from the heart.

Next, that this flux comes from the heart, and

that it flows by a passage made through the heart out of the great veins, seeing below the ligature the blood enters by the arteries, not by the veins, and the arteries at no time receive blood out of the veins, unless it be out of the left ventricle of the heart. Nor could there any otherwise so great abundance be exhausted out of one vein, making a ligature above, especially so forcibly, so abundantly, so easily, so suddenly, unless the consequences were atchieved by the force and impulsion of the heart, as is said.

And if these things be so, we may very openly make a computation of the quantity, and argue concerning the motion of the blood. For if any one (the blood breaking out according to its usual effusion and force) suffer it to come so for half an hour, no body needs doubt but that the greatest part of it being exhausted, faintings and soundings would follow, and not only the arteries, but the greatest veins would be likewise emptied: Therefore it stands with reason, that in the space of that half hour there passes so much out of the great vein through the heart into the aorta. Further, if you should reckon how many ounces flow through one arm, or how many ounces are thrust within the gentle ligature in 20 or 30 pulsations, truly it would minister occasion of thinking how much may pass through the other arm, both the leggs and both the coluses, and through all the other arteries and veins of the body: and that the flux which is made through the lungs

CHAPTER XII

and the ventricles of the heart, must continually
furnish of necessity new blood, and so make a circuit
about the veins, since so great a quantitie cannot be
furnished from those things we eat, and that it is
far greater than is convenient for the nutrition of
the parts.

It is to be observ'd further, that in the administration of Phlebotomie this truth chances sometime to
be confirm'd; for though you tie the right arm, and
lance it as it should be with a convenient orifice and
administer all things as they ought to be, Yet if fear,
or any other cause, or sounding do intervene through
passion of the mind, so that the heart do beat more
faintly, the blood will by no means pass through but
drop after drop, especially if the ligature be made a
little streighter. The reason is, because the pulse
being but faint, and the out-driving force being but
weak, the enfeebled part is not able to open the
passage and thrust out the blood beyond the ligature,
yea nor to draw it through the lungs, or to remove
it plentifully out of the veins into the arteries. So
after the same manner does it come to pass that
Women's flowers and all other fluxes of blood are
stop'd. This likewise appears by the contrary, for
fear being remov'd, and the spirit recollected, when
they do return to themselves, the pulsifick strength
being now increased, you shall streightway see the
arteries beat more vehemently in that part where
they are bound, and move in the wrist, and the
blood leap out farther through the orifice.

CHAPTER XIII.

The third supposition is confirm'd, and that there is a circulation of the blood from the third supposition.

HItherto concerning the quantitie of blood which passes through the lungs and heart in the centre of the body, and likewise from the arteries into the veins and habit of the body. It remains that we do explain which way the blood flowes back from the extremities through the veins into the heart and how the veins are the vessels that carry it from the extremities to the centre, by which means we think those three grounds propounded will be true, clear, firm, and sufficient to gain credit.

But this shall be plain enough from the portals which are found in the concavities of the veins, their use, and from ocular experiments.

The most famous *Hieronimus Fabricius ab Aquapendente*, a most learned Anatomist, and a venerable old man (or, as the most learned *Riolanus* would have have it, *Jacobus Silvius*) did first of any delineate the membranal portals in the veins being in the figure of a Σ, or semilunarie, the most eminent and thinnest parts of the inward tunicles of the veins: Their situation is in different places, after a various manner, in diverse persons they are connate at the sides of the veins, looking upwards towards the roots of them, and in the middle capacity both of them (for they are for the most part two) looking towards one another equally and duly touching one another, insomuch that they are apt to stick together at the extremities, and to be joynd: and lest they should hinder any thing to return from the roots of the veins into the little branches, or from the greater into the less, they are so plac't that the horns of the hindermost are stretched towards the middles of the body of it which is before, and so interchangeable.

The finder out of these portals did not understand the use of them, nor others who have said lest the blood by its own weight should fall downward: for there are in the jugular vein those that look downwards and do hinder the blood to be carried upwards. I (as likewise others) have found in the emulgent veins and branches of the Mesenterie, those which did look towards the *vena cava* and *vena porta*; add to this moreover that there are no such in the arteries, and it is to be observ'd that dogs and cattle have all their portals in the dividing of the crural veins at the beginning of the *os sacrum*, or in the Iliac branches near the *Coxendix*, in which there is no such thing to be feared by reason of the upright stature in man. Nor are their portals in the jugulars, as others say, for fear of Apoplexie, because the matter is apt in sleep to flow into the head through the soprall arteries.

Nor that the blood may stand still in divarications,

and that the whole blood should not break in into the small branches or those which are more capacious: for they are likewise plac'd where there are no divarications, though I confess they are more frequent where divarications are.

Nor that the motion of the blood may be retarded from the centre of the body; for it is likely that it is thrust leysurely enough of its own accord, out of the greater into the lesser branches, and so that it is separated from the mass and fountain. But the Portals were meerly made, lest the blood should move from the greater veins into the lesser and tear or swell them; and that it should not go from the centre of the body to the extremities, but rather from the extremities to the centre. Therefore by this motion the small Portals are easily shut, and hinder any thing which is contrary to them; for they are so plac'd and ordain'd, that if any thing should not be sufficiently hindred in the passage by the hornes of the formost, but should escape as it were through a chinck, the convexity or vault of the next might receive it, and so hinder it from passing any further.

I have often tried that in dissection if beginning at the roots of the veins I did put in the Probe towards the small branches with all the skill I could, that it could not be further driven by reason of the hinderance of the Portals: On the contrary, if I did put it outwardly from the branches towards the root, it passed very easily. In many places two Portals are so interchangeably plac'd and fitted, that

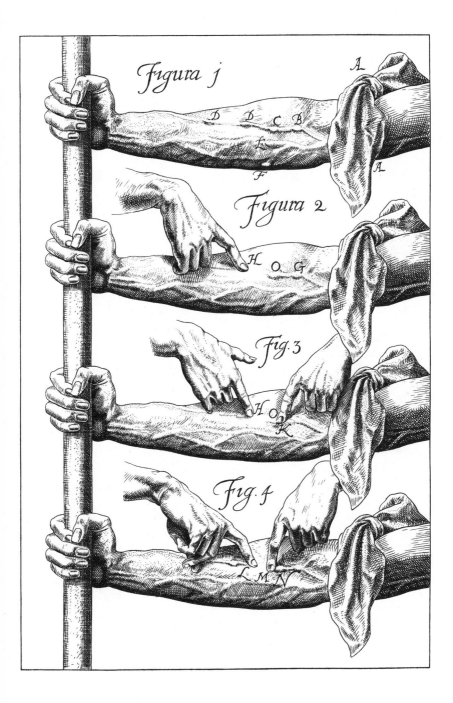

when they are elevated in the middle of the concavity of the vein, they close with one another to a hair's bredth, and in their extremities and convexities are united interchangeably that you can neither see with your eye-sight nor any way discern any crevice or conjunction: on the contrary from outwardly putting in a Probe they easily give way (and like those gates or sluces by which the course of rivers is stopt) they are easily turn'd back to intercept the motion of the blood from the *vena cava* and the heart, and being closely lifted up in many places whilst they are interchangeably shut they do quite hinder and suppress, nor by any means suffer the blood to move neither upwards to the head nor downwards to the feet, nor to the sides or arms, but do stop and resist all manner of motion of the blood, which is begun in the greater veins and ends in the lesser, yet do obey any which is begun by the small veins and ends in the greater, and does provide a free and open way for it.

But that this truth may the more clearly appear, let the arm of a man alive be tied above the Elbow, as if it were to let blood (A A); there will appear at intervals, especially in country people and those who are swoln vein'd, certain little nodes or swellings (B C D D E F), not only where the divarication is (E F), but likewise where there is none (C D), and these nodes are made by the portals. They thus appearing in the inside of the hand or cubit, if you draw down blood with your thumb or finger from the node O to H in the second figure, you shall see that none can follow (the portal quite hindring it) and that the part of the vein H O ot the second figure, drawn down betwixt the swelling and the finger, is quite obliterated, and yet full enough above the knot or portal (O G). Nay, if you do retain the blood so drove down and the blood emptied (H), and do press downward with t'other hand the upper part of the vein O in the third figure, being full, you shall find that by no means it can be forc'd or driven beyond the portal O; but how much the more you do endeavour to do this, so much the more shall you see at the portal or swelling of O, of the third, the vein swoln and distended, and yet that H O of the third figure is emptie below.

Hence, since a man may make experiment in many places, it appears that the function of the portal in the veins is the same as that of the Sigmoides, or three pointed portals, which are made in the orifice of the aorta or *vena arteriosa*, to wit that they may be closely shut up, lest they should hinder the blood to return back again.

Besides tying the arm again as before (A A), and the veins swelling, if you hold the vein below any swelling or portal at any distance L of the fourth figure, and afterwards with your finger M drive the blood upwards above the portal N, you shall see that part of the vein (L N) to remain empty, and that it cannot return by reason of the portal (H O 2), but taking away your finger (H 3, or L in the fourth figure), you shall see't again fill'd by the lower veins, and be like D C of the 1, so that from hence it appears plainly, that the blood does move towards the upper parts and the heart in the veins, and not on the contrary; and albeit in some places which are not closely shut, or where there is but one portal, the passage of the blood from the centre seems not to be quite hindred, yet for the most part it appears so, or at least that which is negligently perform'd in some places is recompens'd by the portals, in order following, either through their number, diligence, or some other way, insomuch as the veins are the open and patent wayes of returning the blood to the heart, but quite stop'd in its going out from thence.

This is moreover to be observ'd, tying the arm as before, and the veins swelling, and nodes or Portals appearing, if below any Portal in any place where you find the next you place your finger, which may hold the vein, that no blood may go from your hand upwards, then squeeze with your finger the blood from that part of the vein L N above the Portal as was said before, then taking away your finger L suffer it to be fill'd up by those under (as D C) and then pressing again with your thumb in the same place, squeeze out the blood (L N and H O), and do this a thousand times in a little space.

Now if you reckon the business, how much by one compression moves upwards by suppression of the portal, and multiplying that by thousands, you shall find so much blood pass'd by this means through a little part of a vein, that you will find your self perfectly perswaded concerning the circulation of the blood, and of its swift motion.

But lest you should say, that by this means Nature is forc'd, if you do this in portals far distant, and do observe, taking away your thumb, how soon and how swiftly the blood returns and fills the lower part of the vein, I do not doubt but you will find the very same.

CHAPTER XIV.

The Conclusion of the demonstration of the circulation of the blood.

NOW then in the last place we may bring our opinion, concerning the circulation of the blood, and propound it to all men.

Seeing it is confirm'd by reasons and ocular experiments, that the blood does pass through the lungs and heart by the pulse of the ventricles, and is driven in and sent into the whole body, and does creep into the veins and porosities of the flesh, and through them returns from the little veins into the greater, from the circumference to the centre, from whence it comes at last into the *vena cava*, and into the ear of the heart in so great abundance, with so great flux and reflux, from hence through the arteries thither, from thence through the veins hither back again, so that it cannot be furnished by those things which we do take in, and in a far greater abundance than is competent for nourishment: It must be of necessity concluded that the blood is driven into a round by a circular motion in creatures, and that it moves perpetually; and hence does arise the action and function of the heart, which by pulsation it performs; and lastly, that the motion and pulsation of the heart is the only cause.

CHAPTER XV.

The circulation of the blood is confirm'd by probable reasons.

BUT it will not be amiss likewise to add this, that according to some common reasons it is convenient, and it ought to be so. First (*Aristotle de Respiratione, & lib.* 2, 3. *of the parts of creatures & elsewhere*) seeing death is a corruption which befalls by reason of the defect of heat, and all things which are hot being alive, are cold when they die, there must needs be a place and beginning of heat (as it were a Fire and dwelling house), by which the nursery of Nature, and the first beginnings of inbred fire may be contain'd and preserv'd; from whence heat and life may flow, as from their beginnings, into all parts; whither the aliment of it should come, and on which all nutrition and vegetation should depend.

And that this place is the heart, from whence is the beginning of life, I would have no body to doubt.

There is therefore a motion requir'd to the blood, and such a one as that it may return again to the heart; for being sent far away into the outward parts of the body (as *Aristotle* 2, *de partibus Animalium*) from its own fountain, it would congeal and be immovable. (For we do see, that by motion, heat and spirit is ingender'd and preserv'd in all things, and by want of it vanishes.) Seeing therefore, that the blood staying in the outward parts is congealed by the cold of the extremities and of the ambient air, and is destitute of spirits, as it is in dead things, it was needful it should resume and redintegrate, by its return again, as well heats as spirit, and indeed its own preservation, from its own fountain and beginning.

We see, that by the exteriour cold, the extremities are sometimes chill, insomuch as nose, hands, and cheeks do look blew, like those of dead men, because that the blood stands still in them (as it does in carkasses in those parts which are down tending); whence it comes that the members are numm'd and hardly moveable, so that they seem quite almost to have lost life. They could certainly by no means (especially so soon) recover heat and colour and life, unless they were by a new original, a Flux and appulsion of heat, again cherish'd. For how can they attract in whom heat and life are almost extinct? or those that have their passages condens'd and stopt with congeal'd blood, how could they receive the coming nourishment and blood unless they did dismiss that which they before contain'd, and unless the heart were really that beginning from whence heat and life (as *Aristotle, de Respiratione*, 2) and from whence new blood being passed through the arteries imbued with spirit, that which is enfeebled and chill'd might be driven out, and all the parts might redintegrate their languishing heat and vital nourishment almost extinct?

Hence it is that it may come to pass, that the heart being untouch'd, life may be restor'd to the rest of the parts, and soundness recover'd; but the heart being refrigerated or affected with some heavy disease, the whole animal must needs suffer, and fall to corruption. When the beginning is corrupted, (as *Aristotle, 3, de partibus Animalium*) there is nothing which can afford help to it, or those things which do depend upon it.

And hence perchance the reason may be drawn, why in those that with grief, love, cares, and the like are possessed, a consumption or continuation happens, or cacochymie, or abundance of crudities, which cause all diseases and kill men. For every passion of the mind which troubles men's spirits, either with grief, joy, hope, or anxiety, and gets access to the heart, there makes it to change from its natural constitution, by distemperature, pulsation, and the rest, that infecting all the nourishment and weakning the strength, it ought not at all to seem wonderful if it afterwards beget divers sorts of incurable diseases in the members and in the body, seeing the whole body in that case is afflicted by the corruption of the nourishment, and defect of the native warmth.

Besides all this, seeing all creatures live by nourishment inwardly concocted, it is necessary that the concoction and distribution be perfect, and for that cause the place and receptacle where the nourishment is perfected, and from whence it is deriv'd to every member. But this place is the heart, since it alone of all the parts (though it has for its private use the coronal vein and arterie) does contain in its concavities, as in cisterns, or a celler (to wit ears or ventricles), blood for the publick use of the body; but the rest of the parts have it only in vessels for their own behoof, and for private use. Besides, the heart only is so plac'd and appointed, that from thence by its pulse it may equally distribute and dispence (and that according to measure and the concavities of the arteries, which are to supply every part) to those which want, and deal it after this manner, as out of a treasure and fountain. Moreover to this distribution and motion of the blood, violence, and an impulsor is requir'd, such as the heart is. To this add, that the blood does easily concentricate and joyn of its own accord to its beginning, as a part to the whole, or as a drop of water spilt upon the table to the whole mass, as it does very swiftly for slender causes, such as are cold, fear, horror, and the like. Besides, it is squeez'd out of the capular veins into the little branches, and from thence into the greater, by the motion of the members and muscles: Likewise the blood is apter to move from the circumference to the centre than otherwise, though the portals did not hinder. From whence it follows, that if it do leave its beginning, and move against its will, and enter into places narrower and colder, that it has need of violence and an impulser, such is the heart only, as we said but now.

CHAPTER XVI.

The circulation of the blood is prov'd by consequence.

THERE are likewise Questions, which, from this supposed verity, for creating of belief, as arguments *à posteriore*, are not altogether unuseful. These though they be envelop'd in much doubtfulness and obscurity, yet easily admit of the assignation of causes and reasons.

We see in contagion, in poisoned wounds, or in the bitings of Serpents, or mad doggs, in the French Pox, and the like, that the part touched being not hurt, it so falls out that the whole habit of the body is vitiated. The French Pox sometimes bewrays it self by the pain of the head, or the shoulders, or other Symptoms, the genitals having no hurt at all. The wound made by the biting of a mad dogg being cured, we have notwithstanding observed, that a feaver and other horrible Symptoms have ensued: Because the contagion being imprinted into the part, it appears that it is from hence carried to the heart with the blood returning, and can afterwards infect the whole body. In the beginning of a tertian feaver the morbifick cause going to the heart makes them breathless, sighing, and lazie, because the vital beginning is oppressed, and the blood is driven against the lungs and thickned, and finds no passage (I speak this, having had experience from the dissection of them that have dyed in the beginning of the accession) then the pulsations are always frequent, little, and sometimes disorderly: But the heat being increas'd, and the matter obtenuated, the wayes being open and passages made, the whole body grows hot, the pulses become greater and more vehement, the Paroxism of the feaver growing higher, to wit, the preternatural heat being kindled in the heart, is diffus'd from thence by the arteries into the whole body, together with the morbifick matter, which by this means is overcome and dissolved by nature.

Likewise, seeing medicaments, outwardly applyed, ever use their force within, as if they were taken outwardly: (Coloquintida and Aloes loosen the belly; Garlick applyed to the soles of the feet, causes expectoration; Cantharides move urine, and cordials do corroborate, and infinite of this kind); From hence it is constantly averr'd, perchance not without cause, that the veins, through their orifices, draw a little of those things which are outwardly applyed, and carry it in with the blood, after the same manner as those in the Mesenterie do suck the Chylus out of the intestines, and carry it to the liver, together with the blood.

In the Mesenterie likewise, the blood entering into the Cœliac arterie, the upper and neather Mesenterics goes forward to the intestines; by which, together with the Chylus attracted by the veins, it returns through the many branches of them into the *Porta* of the liver, and through it into the *vena cava*; so it comes to pass, that the blood in these veins is imbued with the same colour and consistence as in the rest, otherwise than many believe: for we must needs believe, that it very fitly and probably comes to pass, in the stem or branch of the capular veins, that there are two motions, one of the Chylus upwards, another of the blood downwards; but is not this done by a main providence of nature? for if the raw Chylus should be mix'd with the concocted blood in equal proportions, no concoction, transmutation, or sanguinification should from thence arise: But rather (since they are interchangeably active and passive) from the union of them being altered there should arise a mixture, and a thing of a middle nature betwixt the two (as in the mixing of wine and water there is begotten a wine-foyl). But now, when with the great quantity of blood which passes by, a part of the Chylus is mix'd after this manner, and as it were in no remarkable proportion, that doth (as *Aristotle* says) more easily come to pass; as when one drop of water is put into a Hogshead of wine, or on the contrary, the whole is not mixed, but it is either wine or water; so in the Messeraick veins, being dissected, there is found a Chylus, not the Chylus and blood a part, but mixed, and the same both in colour and consistence to the sense, as appears in the rest of the veins; in which notwithstanding, because there is something of the Chylus inconcocted, although insensible, Nature hath placed the liver, in the Meanders or crooks of which it is delay'd, and receives a fuller transmutation; lest coming too soon raw to the heart, it should overwhelm the beginning of life. Hence in Embryons there is no use of the liver where the Umbilical vein doth apparently pass through the whole, for there stands out of the porta of the liver a hole or Anastomosis, that the blood returning from the intestins of the birth, passing not through the liver, but the forementioned Umbilical vein, might go to the heart, together with the mother's blood returning from the Placenta of the womb; from whence likewise, in the first forming of the birth, it comes to pass, that the liver is made last. We likewise in a woman's untimely birth, have observ'd all the members shap'd, the Genitals distinctly, and yet scarce any foundation of the liver to have been laid. And truly so long as the members (as likewise the heart it self in the beginning) are all whole, and that there is no rednesse conteyn'd in the veins, you shall see nothing but in rude collection, as it were of blood without the vessels, instead of the liver, which you would think to be some bruse or broken veins.

There are in an Egg as it were two Umbilical vessels, one passing through the whole liver from the white and going directly to the heart; the other going from the yolk and ending in the *vena porta*. For so it is, that a Chick is first only nourished and found by the white, and afterwards by the yolk, after its perfection and exclusion; for the yolk may be found to be contein'd in the belly of the Chick many dayes after the hatching, and it is answerable to the nourishing of milk in other creatures. But we shall speak of these things more conveniently in our observations concerning the forming of births, where there may be many enquiries of this nature, why this is first made and perfected, and that afterwards; and of the principalitie of Members, what part is the cause of another; and many things likewise concerning the heart, As why (as *Aristotle, de partibus Animalium* 3) it was made the first consistent and seems to have in it life, motition, and sense, before any thing of the rest of the body be perfected: And likewise of the blood, why before all things, and how it has in it the beginning of life and of the creature; why it requires to be mov'd and driven up and down; and then for what cause the heart seems to have been made.

After the same manner in the speculation of pulses, to wit, why such are deadly, other not; and in all kinds by contemplation of their causes and presages, what those signifie, and what these, and why.

Likewise in the crisises and expurgations of Nature; in nutrition, especially in distribution of the nutriment; and likewise in all fluxions, &c.

Lastly, in all parts of Physick, Physiological, Pathological, Semeiotick, Therapeutick, when I do consider with my self how many questions may be determined, this truth and light being given; how many doubts may be solved, how many obscure things made clear, I find a most large field, where I might run out so far and enlarge my self so much, that it would not only swell into a great volume, which is not my intention, but even my lifetime would be too short to make an end of it.

Therefore in this place, that is to say, in the following Chapter, I shall onely endeavour to refer those things to their proper uses and causes, which do appear in the administration of Anatomie, about the fabrick of the heart and arteries: for there where I intend to address my self, very many things are found which receive light from this truth, and do in return make it more clear, which I desire to adorn and confirm by Anatomical arguments beyond all the rest.

There is one thing, which although it ought to have place too in our observations concerning the use of the Milt, yet will it not be impertinent to take notice of it here by the by.

From the splenick veins drawn down into the Pancreas, there arise veins from the upper part of it: the Coronal, Postick, Gastrick, and Gastro-epiploick; all of which, with very many branches and tendons, are dispers'd into the ventricle, as the meseraicks are into the intestines. Likewise from the inferior part of this splenick, down as far as the Colon and Longanon, the Haemorrhoidal vein is deducted. The blood returning through those veins by both ways, and carrying the rawest juice with it, (hence from the ventricle, that which is waterish and thin, the chilification being not as yet perfected; from thence that which is gross and terrestrial) in this branch of the splenick, by the permixtion of contraries, it is conveniently temper'd; and Nature mixing those two juices of more difficult concoction, by reason of their contrary indispositions, with great abundance of warm blood, which (by reason of the abundance of arteries) flows abundantly from the milt, it brings them, being now better prepar'd, to the *porta* of the liver, and supplies and recompences the defect of both by such a structure of the veins.

CHAPTER XVII.

The motion and circulation of the blood is confirm'd by those things which appear in the heart, & from those which appear in Anatomical dissection.

I do not find the heart in all creatures to be a distinct and separate part; for some, as you would say Plant-animals, have no heart; Colder creatures of a softer make, and of a kind of similarie constitution, such as are Palmer-worms and Snails, and very many things which are ingender'd of putrefaction and keep not a species, have no heart, as needing no impulsor to drive the nutriment into the extremities: For they have a body connate and of one piece, and indistinct without members; so that by the contraction and returning of their whole bodie, they take in, expell, move and remove the nourishment, being call'd Plant-animals; such as are Oysters, Mussles, Sponges, and all sorts of Zoophytes, have no heart; for instead thereof they use their whole body, and this whole creature is as a heart.

In very many, and almost all kinds of Insects, by reason of the smallness of their Corpulency, we cannot rightly discern; yet in Bees, flies and wasps we may by the help of a perspective glass. You may likewise see something beat in lice, in which moreover you may clearly see the passage of the nourishment through the intestines (this Animal being transparent) like a black spot, by help of this multiplying glass. But in those that have no blood and are colder, as in Snails, Shell-fish, Crusted-Shrimps, and the like, there is a little part which beats (like a little bladder, or an ear) without a heart, making its contraction and pulse seldomer, and such a one as you cannot discern but in summer, or in a hot season.

In these creatures this particle is ordain'd too that there is a necessity of some impulsion for the distribution of the nourishment, by reason of the variety of the organick parts, or the thickness of their substance: but the pulsations are made seldomer, sometimes not at all, by reason of their coldnesses, as it is meetest for them, being of a doubtful nature, so that sometimes they seem to live, sometimes to die, and sometimes to live the life of an animal, sometimes the life of a Plant.

This is likewise contingent to those Insects which do lurk in the Winter, and are hid as if they were dead, and do only lead the life of a Plant; but whether this do likewise happen to some creatures that have blood, as to Frogs, Snayls, Serpents, Swallows, we may not without reason make a question.

In creatures which are a little bigger and hotter, as having blood in them, there is an impulsion of the nutriment requir'd, and such a one perchance as is endued with more force; therefore in Fishes, Serpents, Snakes, Snails, Frogs, and others of the like nature, there is both one ear and one ventricle of the heart allotted, whence rises that most true Axiom of *Aristotle, de partibus Animalium* 3, That no creature having blood does want a heart, by the impulsion of which it is made stronger and more robust, and the nutriment is not only stirr'd up and down by the ear, but likewise is thrust out further and more swiftly.

That in creatures yet greater, hotter, and more perfect, (as abounding with a great deal of hotter blood and full of spirit) there is a stronger and more fleshie heart requir'd, that the more strongly, more swiftly, or with greater force the nutriment may be thrust out, by reason of the bigness of the body and thickness of the habit.

And moreover, because that more perfect creatures need more perfect aliment and a more abundant native heat, that the nutriment of them may be concocted and acquire a further perfection, it was fit that these creatures should have lungs and another ventricle, which should drive the nutriment through them.

So in whatsoever creature there is lungs, there is likewise in them two ventricles of the heart, the right and the left, and wheresoever the right ear is in any, there is the left, not on the contrary, that where the left is, there is the right one too; that I call the left ventricle which is distinguished in place, but not in use from the tother, which doth diffuse the blood into the whole body, not into the lungs alone, hence the left ventricle seems to make up the heart of it self, being placed in the middle, and so fenc'd with higher ditches, and fram'd with greater diligence that the heart seems to have been made for the left ventricle's sake, and the right ventricle seems as it were a servant to the left, and does not reach to the top of it, and is made up of a thinner threefold wall, and it has, as *Aristotle* says, a kind of articulation above the left, and is more capacious, as administring not only matter to the left, but giving nourishment likewise to the lungs.

But it is to be observ'd in Embryons these are far otherwise, and that there is no such great difference of the ventricles, but like two kernels in a nut they are almost equal, the corner of the right reaches the top of the left, so that in them the heart hath as it were a double top at the point. These things come to pass because in them whilst the blood does not pass through the lungs, as it does pass from the right bosome of the heart to the left, both the ventricles do perform alike the office, bringing the blood through from the *vena cava* into the *arteria magna* by that oval hole and arterious passage, as hath been said, and do equally divide it into the whole body, whence proceeds an equal constitution. But when it is time that the lungs should be used, and the foresaid unions begin to be stop'd, then does this difference of ventricles begin to be in their strength, as likewise in the rest, because the right drives only through the lungs, the left through the whole body.

There are, besides these, in the heart also tendons, as I may so say, or fleshie twigs, and very many fibrous connexions, which *Aristotle* in his book *de Respiratione* and *de partibus Animalium* 3 calls nerves, of which some apart are stretch'd with divers motions, and are partly hidden in furrows with deep ditches about them in the walls and mediastin, and they are like a kind of little muscles which are underordained and superadded to the heart as auxiliaries for the further expulsion of blood, that, like the diligent and artificial provision of tackling in a ship, they might help the heart contracting it self every way, and might squeeze out the blood more fully and forcibly out of the ventricles.

And this is manifest from hence, because some animals have them, some not, and all which have them are stronger in the left ventricle than in the right; some animals have them in the left, and not at all in the right; in men there are more of them in the left than in the right, and more in the ventricles than in the ears, and in some ears almost none; there are more of them in brawnie, musculous and rural bodies, and such as are of rougher habit of body than in those which are tender, and in women there are fewer.

In those creatures in which the ventricles within are smooth, altogether without fibers and tendons, and which are not cleft into ditches (as almost in all little birds, Serpents, Frogs, Snails, and the like, in the Partridge likewise and the Hen, and the greatest part of Fishes) in them neither those nerves or fibers mentioned, nor the three-fork'd portals are to be found in the ventricles. In some animals the right ventricle is smooth within, the left has those fibrous connexions, as in the Goose, Swan, and greater birds: In them the same cause is alleged as in all, seeing their lungs are spongious and soft they need no such force to impell the blood through them; therefore in the right ventricle either they have no fibers, or else fewer and weaker, nor are so fleshy and comparable to Muscles, but in the left they are stronger and more in number, more fleshy and musculous, because the left ventricle hath need of more strength and force, by reason that it ought to pursue the blood farther through the whole body.

From hence it is likewise, that the left ventricle possesses the middle of the heart, and hath a wall threefold thicker, and is stronger than the right ventricle. Hence all creatures, men likewise, by how much the habit of their flesh is harder and more solid, and by how much more their outward members are more fleshy, and farthest from the heart, and brawnie, so much more fibrous, thick, robust, and musculous a heart have they, and this is necessary and clear; on the contrary, by how much the more they are fine-spun, of a softer habit, and of slenderer bodies, so much the softer, flagging, and less fibrous heart within (or not at all) have they.

Likewise consider the use of the portals, which

were made for that cause, lest the blood once let out should be returned to the heart, and as well in the orifice of the arterie, as of a vein, they are up-lifted, and enterchangeably joyning, they make a three square line, such as is imprinted by the biting of a Swallow, that being shut more closely they may hinder the reflux of blood.

There are three forked portals in the entry of the *vena cava* and *arteria venosa*, lest that when the blood is most driven out it should fall back, and for that cause they are not in all creatures, and in those in which they are, they do not seem to be made by the same diligence of Nature, but in some they are shut more exactly, in others more care-lessly and negligently; therefore in the left ventricle, that for the greater impulsion there may be a closer stoppage, there are only two like a Mitre, having tendons reaching out far, even to the *conus* of it, through its middle, that they may be most exactly shut. This perchance deceiv'd *Aristotle*, in making him believe that this ventricle was double, the divi-sion being made athwart, lest the blood should fall back again into the arterie, and by that means the strength of the left ventricle in driving forth the blood into the whole body should be destroyed, therefore these portals do much surpass in bigness, strength, and exact shutting, those which are placed in the right. Hence likewise of necessity, no heart is seen without a ventricle, since it ought to be the well-spring, fountain, and cellar of blood. The same does not always happen in the brain; for almost all sorts of birds have no ventricle in the brain, as it appears in the Goose and Swan, the brains of these, although the brains of a Conie be almost as big, yet the Conie hath ventricles in the brain, the Goose has not. Likewise, wherever there is one ventricle, there hangs by it an ear flagging, cuticular, hollow within, full of blood; where there are two ventricles, there are likewise two ears; on the contrary, there is only one ear in some creatures, or at least a bladder answerable to an ear, or the vein it self dilated (but not the ventricle of the heart) making a pulse instead of the heart, as it appears in Hornets, Bees, and other Insects, whom I believe I can demonstrate by some experiments, to have not only a pulse but a respiration likewise in that place which they call the tail; whence it happens that it is lengthened and contracted, sometimes oftner, sometimes more seldome, according as they seem more panting or to be more indigent of air; but of this in the treatise of Respiration. It is likewise manifest that the ears do beat and contract themselves, as I said before, and cast the blood into the ventricle, whence it is that wheresoever there is a ventricle there an ear is requir'd, not only (as is commonly believ'd) that it may be the receptacle and cellar of blood (for what needs there any pulsation for the retaining of it?), but the first movers of the blood are the ears, especially the right, being the first thing that lives, and the last that dies, as before is said; for which cause they are necessary, that they may serve to pour the blood into the ventricle. But the ventricle immediately contracting itself, doth more conveniently squeeze out and more violently thrust forth the blood, being already in motion; as when you play at ball, you can strike it farther and more strongly, taking it à la vole, than you could only throwing it out of your hand. But likewise, contrary to the vulgar opinion, because neither the heart, nor any thing else can so extend it self as that it can attract any thing in its diastole (unless in its return to its former constitution, being before squeezed like a spunge); but it is certain, that all local motion comes first, and did take its beginning from the contraction of some particle; therefore by the contraction of the ears, the blood is cast into the ventricle as I open'd before, and by the contraction of the ventricles, it's thrown farther and removed.

Which truth concerning local motion, and that the immediate motive organ (in all creatures in which a motive spirit is primarily) is contractible, as *Aristotle* sayes in his book *de Spiritu*, and elsewhere, and in what way νεῦρον is derived from νεύω, *nuto* from *contraho*, and that *Aristotle* did know the muscles when he did refer all the pains and motion in creatures to the nerves, or that which is contractable, and therefore call'd those tendons in the heart, nerves; I hope it shall be made clear if at any time I shall have liberty to demonstrate concerning the motive organs of creatures and the fabrick of the muscles from my own observations.

But pursuing our purpose concerning the use of the ears, which we did demonstrate was to fill the ventricles with blood, we see it comes to pass, that the thicker and more compaċt the heart is and of a grosser wall, the more nervous and musculous the ears are to draw in and fill it; and in those in whom they are contrarywise, it does appear in them as a bladder of blood, or a membrane conteyning blood, as in fishes, for there the bladder which is in lieu of the ear is very thin, and so large that the heart seems to swim above it; but in those fishes in which this bladder is a little more fleshie, it seems very precisely to emulate and counterfeit the lungs, as in the Barbell, Tench, and others.

In some men, to wit such as are brawnie and of a rougher habit of body, I have found the right ear so strong and so neatly made up within, with the various contexture of fibers, that it did seem to be equal in strength to the ventricles of other men; and truly I did wonder that in divers men there should be such difference. But it is to be observ'd, that in the birth, the ears are far greater than they are in it proportionated, because before the heart is made, or assumes its own funċtion, (as before was shew'd) they do the office of the heart.

But the things that I observ'd concerning the forming of the birth which I made mention of before, and *Aristotle* confirms in an egg, do add a great deal of credit and light to this business; first, whilst the birth is as it were a tender worm, and whilst it is yet (as is usually spoken) in the milk, there is in it a little bladder or bag which beats, and as it were a portion of the umbilical vein; afterwards, when the birth, being shaped, begins to have a stronger corpulency, this little bag becoming more fleshie and robust (changing its constitution) turns into ears, above which the body of the heart begins to spring, as yet executing no publick office; but the birth, when 'tis already form'd, and that the bones are distinct from the flesh, and it is a perfect creature, and that it is felt to have motion, then the heart is both found beating within, and does transfuse the blood as I have said out of the *vena* into the arterie through both the ventricles.

So Nature being perfect and divine, and making nothing in vain, neither gave a heart to any where there was no need, nor made it before there was any use for it, but by the same degrees in the forming of all animals passing through the constitutions of all creatures (as I may say in the egg, Worm, and birth) it acquires its perfection in them all. These things shall be confirm'd elsewhere by many observations in the forming of the birth.

Lastly, *Hippocrates* in his Book *de Corde* did not without reason call it a muscle, seeing the action and and function of both is the same, viz. to contract it self, and move somewhat else, that is, the blood.

Moreover, from the constitution of the fibers and their motive frame, as likewise in the muscles, we may see the action and use of the heart. All Anatomists have observ'd with *Galen*, that the body of the heart is made with several draughts of fibers streight, thwart, and crooked, but in a heart, being boyl'd, the structure of the fibers is found to be otherwayes.

For all the fibers in the walls and in the inclosure are circular, as they are in a Sphincter, but those which are in the tendons stretched out in length, are crooked; so it comes to pass that when all the fibers are contracted, it happens that the top is brought to the bottom by the tendons, and the walls are inclosed in a round, and the heart is contracted every way, and the ventricles strengthned. Wherefore since the action of it is contraction, we must needs imagin that the function of it is to thrust blood out into the arteries.

Nor must we disagree trom *Aristotle* concerning the principality of the heart, that it does not receive motion and sense from the brain, nor blood from the liver, but that it is the beginning of the veins and of the blood, and the like; Seeing those that endeavour to confute him omit that chief argument, to wit, That the heart is the first subsistent, and that it hath blood, life, sense, and motion before the brain or liver were made, or appear'd distinctly, at least before they could perform any function. To this add, That the heart, as a sort of internal animal, consists longer, as if Nature by the making of this first, would have the whole animal afterwards to be made, nourish'd, preserv'd, perfected by it, as its own work and dwelling place. The heart is as it were a Prince in the Commonwealth, in whose person is the first and highest government every where; from which as from the original and foundation, all power in the animal is deriv'd, and doth depend.

But besides, very many things about the arteries do likewise evidence and confirm this truth, when it is consider'd why the *arteria venosa* does not beat since it is numbred amongst the arteries; or why there is a pulse found in *vena arteriosa*, since the pulse of the arteries arises from the impulsion of blood; or that the arteries in the thickness of their tunicles, and the strength of them, do differ so much from the veins, because they bear the force of the impulsion of the heart, and breaking out of the blood.

Hence, since Nature who is perfect, makes nothing in vain, and is sufficient in all things, the nearer the arteries are to the heart, the more they differ from the veins in their constitution, and are more robust and full of ligaments, but in the furthest dispersions of them, in the hand, foot, brain, mesenterie, and spermatick vessels, they are so like in their constitution, that earnestly viewing their tunicles, it is a hard business to know one from the other.

And this is so for just causes. For the further the arteries are distant from the heart, by so much less strength a great deal are they struck, the stroak of the heart being weakned by the great distance. Add to this, that the impulsion of the heart, since it must needs be sufficient in all the trunks and branches of the arteries, it is lessen'd at every partition, as being divided, insomuch that the last divisions of the *capillares arteriosae* seem to be veins, not only in constitution, but likewise in function, or do not give a sensible pulse, or none at all, or else not alwayes, unless the heart do beat more forcibly, or some little arterie be dilated, or more open in some part. Hence it comes, that sometimes we may find a pulse in the teeth, sometimes in the gums, and sometimes we cannot. From hence I did certainly observe, that Boys, whose pulses are alwayes swift and frequent, were in an undoubted Feaver, by this one token; as likewise in tender and delicate people by griping of their fingers, I could easily perceive by the pulse of their fingers when the Feaver was in its strength.

On the other side, when the heart beats faintly, not only not in the fingers, but neither in the wrist, nor in the temples can any pulse be felt, as in fainting, hysterical symptoms, defect of pulse, weak people, and those that are departing.

Here Chirurgions are to be admonish'd, lest they be deceiv'd; because in the cutting off of members, the cutting away of fleshy tumors, and in wounds, the blood does indeed come forcibly out of the arterie, but not alwayes with leaping, and that the small arteries do not beat, especially if they be tyed with a ligature. Beside, that the *vena arteriosa* hath not only the constitution and tunicle of an arterie, but that it does not differ so much in the thickness of the tunicle from the veins as the aorta. The reason is, because the aorta abides a greater impulsion of the blood from the left ventricle, than that does from the right; therefore it has the constitution of the tunicles so much the softer than the aorta, by how much the right ventricle of the heart is weaker than the left; And by how much the contexture and softness of the lungs does abate from the habit of the body and flesh, so much does the tunicles of the *vena arteriosa* differ from that of the aorta.

All these things do constantly keep proportion in men, for the more brawny, musculous, and of harder habit of body they are, and the stronger, thicker and more fibrous heart they have, so much the more answerable ears and arteries proportionably they have in thickness and in strength. Hence in those creatures, the ventricles of whose hearts are smooth within, without roughness, portals, and with a thinner wall, as in Fishes, Birds, Serpents, and very many sorts of creatures, in them the arteries differ very little or nothing from the thickness of the veins.

Besides, the lungs have such large vessels, their vein and arterie, that the trunk of the *arteria venosa* does exceed both the crural and jugular branches, and are so full of blood, as by experience and my own eye-sight (nor was I deceived in the inspection of those things which I saw in dissected creatures) that upon the wounding of them, all the whole blood has run out; the cause, by reason that in the lungs and in the heart is the fountain, cellar, and treasure of blood, and store-house of its perfection.

Likewise we see in Anatomical dissection, that the left ventricle and the *arteria venosa* does abound with so great a quantity of blood, and indeed of the same colour and consistence with that with which the right ventricle and the *vena arteriosa* is fill'd, alike black and clotted, because the blood passes hither from thence continually through the lungs.

Lastly, the vein call'd *arteriosa*, commonly has the constitution of an arterie, the *arteriosa venosa* of a vein, because in truth, both in function, constitution, and all things else, that is an arterie, and this a vein, otherwise than is commonly believed; besides, the *vena arteriosa* hath such a wide orifice, because it carries a great deal more blood than is necessary, for nourishing of the lungs.

All these Phaenomena's to be observ'd in dissection, and very many more, if they be rightly weigh'd, seem to clear the foresaid truth abundantly, and indeed to confirm it, and withall to go against the common opinion: Seeing it is very hard for any to demonstrate by any other way than we have done, for what cause all these things are appointed.

F I N I S.

DE CIRCULATIONE SANGUINIS

TWO ANATOMICAL EXERCITATIONS

CONCERNING

THE CIRCULATION OF THE BLOOD

TO

JOHN RIOLAN

THE SON

THE FIRST ANATOMICAL EXERCITATION CONCERNING

The Circulation of the Blood,

TO

JOHN RIOLAN.

THERE did come forth not many moneths agoe
a little piece of the most famous *Riolan's*
concerning Anatomie and Diseases; for which, as
being sent to me by the Author himself, I return
hearty thanks: Seriously I do congratulate the felic-
ity of that man in undertaking a thing very com-
mendable. To open to the view the seats of all Diseases, is a work not to be atchieved but by a divine
wit. Truly he undertook a hard task, that has set
those Diseases, which are almost obscure to our understanding, before our eyes. Such endeavours become the Prince of Anatomists; for there is no
Science which has not its beginning from foregoing
knowledge, nor any knowledge which is not beholding to sense for its original: for which cause
the business it self, and the example of so worthy
a person required my pains, and did invite me in
like manner to put forth and joyn my medicinal
Anatomie, being chiefly fitted for Physical uses,
not with the same intention as he, by demonstrating
the places of diseases, from the dead bodies of
healthful men, and rehearsing the divers sorts of

diseases incident to those places, according to other men's opinions, which he ought to have seen there; but that I might undertake to relate from the many dissections of sick bodies, and the most grievous and wonderful diseases of dead persons, in what manner and how the inward parts of them are changed in place, bigness, condition, figure, substance, and other sensible accidents, from their natural form and appearance, which all Anatomists commonly described, and how diversly and wonderfully they are affected. For as the dissection of healthful and well habited bodies conduces much to Philosophie and right Physiologie, so the inspection of diseased bodies conduces chiefly to Pathological Philosophie. For the Physiological contemplation of those things which are according to Nature, is first to be known by the Physician, for that which is according to Nature is right, and is rule both to it self and that which is amiss; by the light of which, errors and preternatural diseases being defined, Pathologie is more clear, and from Pathologie the use and art of administring Physick, and occasions of inventing many new remedies do occur. Nor will any man believe how much in diseases, especially such as are Chronical, the inwards are changed, and what monstrous shapes of the inward parts are begotten by diseases: And I dare say the opening and dissection of one consumptive person, or of a body spent with some antient or venemous disease, has more enriched the knowledge of Physick, than the

dissections of ten bodies of men that have been
hanged.

Yet do not I disallow of the most famous and most
learned Anatomist *Riolan* his purpose, but think it
highly to be commended, as being very profitable
for Physick, that he does illustrate the Physiological part; yet did I think that it would not be less
profitable to the art of Physick, if I should set
clearly before your eyes to be seen, not only the
places, but likewise the diseases of those places, and
rehearse them, after I had well viewed and observed
them, and from my many dissections declare my
experience.

But such things in that Book concerning the Circulation of the blood found out by me, which are
translated and seem to reflect only upon me, must
first and chiefly be taken into consideration by me.
For so great a man's judgement concerning such
a weighty business is not to be set at nought (who
is undoubtedly thought the chief and ringleader of
all Anatomists of this age), but the opinion of him
alone is more to be weighed for commendation
than the verdicts of all others, which shall either
applaud or contradict me, and his censure more to
be weighed and looked upon. He then in his *Enchiridion, lib.* 3, *cap.* 8, Acknowledges our motion
of the blood in Animals, and takes part with us,
and is of our opinion as concerning the circulation
of the blood: yet not altogether and openly; for
he says, *lib.* 2, *cap.* 21, That the blood in the port

vein contained, admits no circulation, as the blood in the *vena cava*, and in *lib.* 3, *cap.* 8, That there is blood which is circulated, and circulatory vessels, to wit, the aorta and the *vena cava*, yet he denies that the branches of them have any circulation; *Because*, says he, *the blood running out into all the parts of the second and third region, stays there for nutrition, nor does it flow back to the greater vessels, but being plucked back, by force, when the greater vessels are in great want of blood, or when it returns with a sudden force, or exstimulation to the greater circulatory vessels.* And so a little after: *Whether or no the blood of the veins, does perpetnally or naturally ascend? or whether it returns to the Heart? or whether the blood of the Arteries do descend, or go from the Heart? yet if the lesser veins of the arms and legs be empty, the blood of the veins in succession filling the empty places, may descend, which* (says he) *I have clearly demonstrated against* Harvey *and* Wallaeus. And because daily experience and the authority of *Galen* does confirm the *Anastomosis* of the veins and arteries, and the necessity of the Circulation of the blood; *You see,* says he, *how the circulation of the blood comes about, without the confusion of humours, or the perturbation of antient medicine.*

By which words it is known for what cause the most famous man would partly acknowledge, partly deny the Circulation of the blood and why he endeavours to build a reeling and tottering opinion of Circulation: Lest, forsooth, he should destroy the antient Physick, and not moved by truth which he could not chuse but see, but rather for fear he should violate the antient rules of Physick, or perchance, lest he should seem to resume or retract that Physiologie which in his *Anthropologia* he had publish'd before. For the Circulation of the Blood does not destroy the ancient Physick, but furthers it; rather it shews the Physiologie of Physicians, and the speculation of natural things, and disallows the Anatomical doctrine of the use and action of the heart, lungs, and the rest of the intrals; and that these things are so, will appear partly out of his own words, partly out of those things which I shall here set down; namely, that the whole blood, in whatsoever part of the body living it be, does move and shift place, (as well that which is in the greater veins and their branches and fibers, as that in the porosities of the parts in any region of the body) does flow to the heart and flow from the heart, without interruption, incessantly, and never continues in one place without damage; though I do not say, but in some places it moves slower, in some faster.

First then the most learned man denies only that the blood contained in the *Porta* does circulate which he could neither have denied nor disapproved of, if he had not passed over the force of his own argument: for he says, *lib.* 3, *cap.* 8, *If in every pulsation the heart receive one drop of blood which it expels into the aorta, and does make two thousand pulsations in an hour, there must needs a great deal of blood pass through.*

125

He is likewise forced to affirm the same of the mesenterie, since through the cœliacal arterie and the mesenterial arteries, there is thrust in more than one drop of blood at every pulsation, and is forced against the mesenterie and its veins: insomuch that it must either go out according to the just proportion of that which enters, otherwise the branches of the *Porta* would burst at last; nor can it (for the resolution of this doubt) be probably said, or possibly be, that the blood of the mesenterie should vainly, and to no purpose, ebb and flow through these arteries, like an Euripus; nor the relapse from the mesenterie by those passages and transplantation by which he would have the mesenterie disgorge it self into the aorta, likely to be true; nor can it prevail against that which is entring by contrary motion; nor can there be any vicissitude, where it is most certain that without interruption and incessantly there is an influx; but is compelled by the same necessity, by which it is certain that the heart doth thrust forth the blood against the *mesenterium*. Which is most manifest: for otherwise, by the same argument and subterfuge, he would overthrow all Circulation of the blood, if thus he should, with the same likelihood of truth, affirm that too in ventricles of the heart, namely, in the Systole of the heart the blood is driven into the aorta and in the Diastole returns, and the aorta disburthens it self into the ventricles of the heart, as the ventricles again into the aorta, and so neither in the heart nor in the mesenterie should there be any circulation, but a flux and reflux, by turns, is turned up and down with needless labour: Therefore if of necessity in the heart is proved the circulation of the blood, for the reason aforesaid proved by himself, the same force of argument takes place likewise in the mesenterie; but if there be no circulation in the mesenterie, neither is there in the heart; for both these assertions, namely, this of the heart, that of the mesenterie, hangs upon the force of the same argument, only changing the words, and is established and falls in like manner.

He says, that the Sigma-like portals do hinder the regress of the blood in the heart, but there are no portals in the mesenterie.

I answer, neither is this true; for in the splenick branch, as likewise sometimes in others, there are found portals. Besides, portals are not all times requisite in the more profound veins, nor are they found in the deep veins of the joints, but rather in the skin veins; for where the blood flowing out of the less branches is prone naturally to come into the greater, by the compression of the muscles about it, it is sufficiently hindered from return, but where the passage is open, it is forced. What need is there then of portals? But how much blood at every pulsation is forced into the mesenterie, is reckoned according to the same account, as if with an indifferent ligature you should in the carpus bind the veins coming out of the hand and entring into the arteries (for the arteries of the mesenterie are greater than those of the carpus); if you tell at how many pulsations the vessel and your whole hand swell to their greatest bigness, dividing and making a subduction, you shall find much more than one drop of blood come in at every pulsation, notwithstanding the ligature; nor can it return, but rather that in filling the hand it forcibly distends and swells it, we may by calculation gather, that the blood enters the mesenterie in the same quantity, if not in a greater, by how much the arteries of the mesenteries are greater than those of the carpus. And if any should but see and think with himself, with what difficulty and pains, compressions, ligatures, and several means the blood is stayed, that leaps forcibly out of the least arterie which is cut or broken, with what strength (as if it were shot out of a spout) it throws off, and drives away, or passes through all the bindings, I think he would scarce believe that any part of blood which only enters, could against this impulsion and influx pass back again, being not able to drive it back with force. For which cause, considering these things with himself, I believe it would not ever enter his mind to imagin that the blood out of the veins of the *porta* could creep back by these same ways; and so disburthen it self into the Mesenterie, against so forcible and strong an influx into the arteries.

Moreover, if the most learned man believe not that the blood is moved and changed by circular motion, but being still the same, it stands and mantles in the branches of the mesenterie; he seems to suppose that there is a two-fold blood, divers, and serving to divers uses and ends, and therefore it is of divers natures in the *vena porta* and *cava*, because one of them for its preservation needs circulation, the other needs not, which neither does it appear, nor does he demonstrate it to be true.

Besides, the most learned man adds in his *Enchiridion, lib.* 2, *cap.* 18, *A fourth sort of vessels to the Mesenterie, which are called the Venae Lacteae* (invented by *Asselius*) *which being set down, he seems to infer that all the nutriment being drawn through them is carried to the liver, the forge of blood, which being there concocted and changed into blood* (he says in *lib.* 3, *cap.* 8) *it is carried to the left ventricle of the heart, which being granted,* says he, *all the scruples which were antiently motioned concerning the distribution of the Chylus, and of the blood throughout the same conduit, do cease, for the Venae Lacteae carry the Chylus to the Liver, and therefore these conduits are apart, and can be obstructed apart.* But indeed I would fain know how this can be demonstrated to be true; if this milk be transfused and passe into the liver, how shall it get thence through the *cava* into the ventricle of the heart? (Since the most learned man denies that the blood contained in the numerous branches of the *porta* and the liver can pass, that so circulation may be made) but more especially since the blood seems to be a great deal fuller of spirit, and more penetrative than the milk or Chylus, which is contained in these vessels, and is hitherto impelled by the arteries that it may find out some way for its self.

The most learned man makes mention of a certain Treatise of his concerning the Circulation of the blood. I wish I could see it; I might perchance recant.

But if the most learned man thought it more fit to place the circular motion of the blood in the veins of the *porta*, and branches of the *cava*, (as he says in his 3 Book, Chap. 8, *In the veins, the blood does perpetually and naturally ascend or return to the heart, as likewise that which is in all arteries descends and departs from the heart*) I say, I do not see, but upon this position, all difficulties which were objected of old of the distribution of the Chylus and blood, through these same conduits, should likewise cease, that hence forward he should not need to enquire apart for, or to set down vessels for the chylus; seeing as the Umbilical veins do draw their nutritive juice from the liquors of the egg and carries it to the nourishing and augmentation of the Chick whilst it is yet an Embryon, so do the meseraick veins suck the chylus from the intestines, and carry it to the liver, and what hinders us to assert, that it does the like in those of riper age? For all difficulties cease, when there are not two contrary motions supposed in the same vessels; but that we do suppose that there is one continued motion in the meseraicks from the intestines to the Liver.

I shall tell you in another place what is to be thought of the *venae Lacteae*, when I shall speak of milk found in several parts of creatures new born, especially in mankind, for it is found in the mesenterie and all its glandules, as also in the *chymus*, likewise in the arm pits and paps of Children; the Midwives milk out the blood for their health as they believe.

But moreover it pleased the most learned Riolan, not only to deprive the blood contained in the mesenterie of circulation, but also he affirms, that neither the branches of the *vena cava*, or its arterie, or any part of the second or third region admits of circulation, so that only he calls the *vena cava* and the aorta circulatory vessels, for which in his 3 Book, Chap 8, he gives a very faint reason, *Because the blood*, says he, *flowing into all parts of the second and third region remains there for nourishment, nor does it flow back to the greater vessels, unless it be revulsed by the force and want of blood in the greater vessels, or flow back, being stirred with a sudden force, to the circulatory vessels.*

It is indeed of necessity, that the portion which passes into nourishment, should remain, for otherwise it should not nourish unless it be assimilated, and stay there, in lieu of that which is lost, and so become one: but it is not needful, that the whole influx of blood should remain there for the conversion of so little a portion; for every part does not use so much blood for its nourishment, as it contains in its veins, arteries, and porosities, nor is it necessary in his afflux and reflux that it should leave no nourishment within it; wherefore it is not necessary that for nutrition it should all stay, but likewise the most learned man himself, in the very same book in which he affirms this, does seem everywhere almost to affirm the contrary, especially where he sets down the circulation in the brain, and by circulation (says he) the brain does send back blood to the heart, and so the heart is refrigerated. After which sort likewise, the remote parts may be said to refrigerate the heart, whence also in feavers, when the parts about the heart are grievously scorched and inflamed with feaverish heat, laying naked their joynts, and throwing off the cloaths, sick people endeavour to cool their heart, whilst (as the most learned man affirms of the brain) the blood being refrigerated and allayed of its heat, does then go to the heart through the veins, and does refrigerate it. Whence the most learned man seems to insinuate a kind of necessity, that as from the brains, so there is a circulation from all the parts, otherwise than before he had openly declared. But indeed he cautiously and ambiguously affirms, That the blood does not flow back from the parts of the second and third region, unless, says he, being revuls'd by the force and great want of blood in the bigger vessels, or that it does by a sudden forcible motion flow back to the greater circulatory vessels, which is most true, if these words be understood in a true sense; for by the greater vessels, in which, he says, want causes a reflux, I believe he understands the *vena cava,* or the circulatory veins, not the arteries; for the arteries are never emptied but into the veins or pores of the parts, but they are continually stuffed full by the pulse of the heart. If all the parts did not incessantly refund blood in abundance into the *vena cava,* and the circulatory vessels, out of which the blood very suddenly passes, and hastens to the heart, there would quickly be a great want of blood. Besides that, the blood which is contained in all the parts of the second and third region, by the force of the blood directed and driven by every pulse, is forced out of the pores into the veins, out of the branches into the greater vessels, as likewise by the motion and compression of the parts adjacent; for that which is contained is thrust out by every thing containing it, when it is pressed and streightned: so by the motion of the muscles and the joynts, the branches of the veins passing between, being pressed and streightned, thrust the blood contained in the lesser vessels into the greater.

But it is not to be doubted, that the blood is continually and incessantly driven, and comes with force from the arteries, and never flows back; if it be admitted, that in every pulse all the arteries together are distended by the propulsion of blood, and that the Diastole of the arteries, as the most learned man confesses, is from the Systole of the heart, nor does the blood once gone forth, return into the ventricles of the heart, by reason that the portals are shut; if (I say) the most learned man does believe these things, as it seems he does, it will easily be understood in every part of what region soever by what stuffing or impulsion the blood in them contained is forcibly thrust down.

For so far as the arteries beat, so far reaches the influx and the force, wherefore it is felt in all parts of every region, for there is a pulse every where in the tops of our fingers, and under the nails, nor is there any part in our whole body, either sore with boil or fellon, which does not feel the pricking motion of the beating of the arterie, and its endeavour to dissolve the *continuum*.

But further, it is manifest, that the blood does make a regress in the pores of the parts, in the skin of the hands and feet, for sometimes in great frost and cold seasons we see the hands and joints, especially of boys, so cold, that at the very touch they do almost resemble the coldness of Ice, and are so benummed and stiff, that there is scarce any life in them nor motion, and yet in the mean time they are full of blood, seeming red or blew, which parts can again by no means be warmed, unless by Circulation that refrigerate blood be thrust out, and in its place, new, warm, and spirituous blood flowing in do foment and re-warm the parts, and restore to them motion and sense; for they should never be renewed or restored by external heat, no more than the members of dead persons, unless some internal influent warmth did refresh them. This indeed is the chief use and end of the Circulation of the blood, for which cause the blood, by its continual course and perpetual influence, is driven about; namely, that all the parts depending upon it by their first innate warm moisture might be retained in life, and in their own vital and vegetative essence, and perform all their funƈtions, whilst (as the Naturalists say) they are sustained and aƈtuated by natural heat, and vital spirits; so by the heat of two extremities, heat and cold, the temper of the bodies of creatures is kept in its mediocrity: for as the breathing in of air does temper the too much heat of the blood in the lungs, and in the centre of the body, and causes the eventilation of suffocating fumes; so also the blood being hot, and cast out through the arteries into the whole body, does foment and nourish the extremities in living creatures, and hinders them to be extinguish'd by the force of outward cold.

Therefore it were injust and wonderful, if every little part of what region soever should not enjoy the benefit of the transmutation and circulation of the blood, for whose sake Circulation seems chiefly to be appointed by Nature. Therefore, that I may conclude, for you see how the Circulation of the blood is performed without perturbation or confusion of the humours in all the body, and in every part, both in the greater and in the lesser vessels, and that by necessity, and for the benefit of all the parts, without which, being cold and impotent, they could never be restored, or remain alive. It is enough, because it's clear, that all influence of preservative heat does come through the arteries, and is done by circulation.

For which cause most learned *Riolan* seems to me, when he says, that in some parts there is no Circulation, to speak rather officiously, than truth; to wit, that he might please most men, and oppose no body, and that he rather wrote humanely, than gravely, in the behalf of the truth. As he likewise seems to do (*lib*. 3, *cap*. 8) when he would rather have the blood to come into the left ventricle through the septum of the heart, through uncertain and hidden passages, than through the large and most open vessels of the lungs, being made with Portals artificially to hinder its return. I desire to see the reason of the impossibility and inconvenience which he says he propounded elsewhere. It is a wonder, since the *Aorta* and *vena Arteriosa* are of the same bigness, constitution, and frame, that their function should not be the same. But that is very improbable that the great River of the whole mass of blood should in so great abundance go into the left ventricle by so blind and small a winding of the septum, which should answer both to the entrie from the *vena cava* in the right side of the heart, and also its egress from the left, which do both require such wide orifices. But he has likewise produced these things staggeringly, for in *lib*. 3, *cap*. 6, he ordains the lungs as a sink or passage from the heart, and he says, *The lungs are affected by that blood which passes through, whilst its filth flows together with that blood*; so he says likewise,

EXERCITATIO PRIMA

That the lungs acquire corruption by distempered, and ill-conditioned intralls, which furnish the heart with impure blood, whose fault the heart cannot help but by many circulations. He likewise in the same place, concerning letting of blood, and shortness of breath, and communication of the veins with the vessels of the lungs, says against *Galen, If it be true that the blood does naturally pass from the right ventricle of the heart to the lungs, that it may be carried to the left ventricle, and so to the aorta; and if the Circulation of the blood be admitted, who sees not in the diseases of the lungs, that the blood flows thither in greater abundance, and oppresses the lungs, unlesse they be first largely emptied, every part taking a share to ease them, which was* Hippocrates *advice from all parts of the body, head, nose, tongue, arms, feet, to take away the blood, that the quantity of it might be impaired, and that it might be revulsed from the lungs, and so draws out the blood till the body was quite without blood.* He says likewise, *The Circulation being supposed, the lungs are easily emptied by breaking a vein. If this counsel be rejected, I see not how it can be revulsed from thence; for if it flow back through the vena arteriosa into the right ventricle, the Sigmoidal portals hinder it, and the three-pointed portals hinder the regress out of the right ventricle into the vena cava. Therefore by Circulation the blood will be exhausted, by cutting the veins of the armes and feet. And likewise* Fernelius *his opinion in the affections of the lungs is destroyed, that blood is rather to be taken out of the right arm than out of the left, for the blood cannot return into the vena cava, unless it break through two gates and bars which are placed in the heart.*

He addes moreover in the same place (*lib. 3. cap. 6.*), *If the Circulation of the blood be admitted, and that it doth pass often through the lungs, and not through the middle of the Septum of the heart, there is a two-fold Circulation of the blood to be assigned, one of which is perfected by the heart and the lungs, whilst the blood leaping out from the right ventricle of the heart is carried through the lungs that it may come to the left ventricle of the heart; for leaping out from the same inward part, it returns to it, then by a another larger circulation flowing out of the left ventricle of the heart, it goes about the whole body, and runs through the arteries and veins to the right ventricle of the heart.*

The most learned man in this place might have added the third circulation, which is a very short one, out of the left ventricle into the right, drawing about a part of the blood through the coronall arteries and veins, by its branches, which are distributed about the bodie, walls, and septum of the heart.

He says, *He that admits of one circulation, cannot deny the other.* So might he have added, nor can he refuse the third. For to what purpose should the coronal arteries beat in the heart, if they did not drive blood thither? and why should the veins, (whose function and end it is to receive blood put into them by the arteries) but that they might draw blood from the heart? Moreover, in the orifice of the Coronal arterie (as the learned man himself confesses in his third Book and his ninth Chapter) there is a portal which forbids all entrance, and is patent to egresse: therefore truely he cannot but admit of the third Circulation, who likewise admits of another universal one, and that the blood does likewise passe through the lungs and the brain (*lib.* 4, *cap.* 2). For neither can there be an admittance of blood by pulsation in all parts of every region, nor regresse by the veins after the same manner, and therefore he cannot deny, but that the parts admit of Circulation.

Therefore it is clear from these very words of the most learned man, what his opinion is, both of the Circulation of the blood through the whole bodie, as likewise through the lungs and the rest of the parts; for he that admits of the first Circulation, it is clear that he does not reject the other: For how can it be that he who has admitted of another Circulation through the whole body so often, and through the greater circulatory vessels, should deny that universal Circulation in any of the branches or parts of the second or third region? As if all the veins and those greater circulatory vessels, as he calls them, were not number'd by himself, and by all others, amongst the vessels of the second region. Is it possible that there should be circulation through the whole body, and not through all the parts? and therefore where he denies it, he does it very stammeringly, and only staggers and palliates in his negations: there where he affirms, he speaks understandingly, and as becomes a Philosopher, and as a skilful

Physician and an honest man gives his advice in this case, that in the dangerous diseases of the lungs, the letting of blood is the only remedy, against *Galen* and his beloved *Fernelius:* in which thing, if he had been doubtfull, far be it from a Christian and so learned a man, to recommend his experiments to posterity, to procure death, and the hazzard of men's lives, or that he should recede from *Fernelius* or *Galen*, men in high esteem with him. Therefore whatsoever he has denyed of the circulation in the mesenterie, or any other part, in favour of the antient Doctrine of Physick, or the *Venæ Lacteæ* or for any other regard, it is to be attributed to his civility and modesty, and to be pardoned.

I think it does already appear clearly enough, both from the words and the arguments of the most learned man himself, that there is a circulation every where, and that blood wheresoever it is, does change place, and passe through the veins to the heart; and the most learned man seems to be of the same opinion with me: Therefore it needs not, yea it were superfluous to bring hither my arguments which I have published in my Book concerning the motion of the blood, for the further confirmation of this truth, which are taken both from the frame of the vessels, placing of the portals, and other experiments and observations; especially since I have not as yet seen the most learned man's Treatise of the Circulation of the blood, nor as yet any of the most learned man's Arguments, but only a bare negation, by which being induced, he should reject the circulation in the regions and vessels, which he allows to be universal in most of the parts.

It is indeed true, that I did find out of the authority of *Galen*, and by dayly experience to be a *refugium* the *Anastomosis* of the vessels, yet so great a man as he is, so diligent, so curious, so expert an Anatomist, should first have laid open and shewn *Anastomoses*, and those visible and open ones, and whirlpools proportionable to the impetuous stream of the whole blood, and the orifices of the branches (from which he has taken away circulation), before he had rejected those which were most probable and most open. He was oblig'd to demonstrate and declare where they are, how they are fram'd, whether they are not only fit for the intromission of blood (as we see the arteries inserted in the bladder) and not for the return of it, or what other way soever they had been. But perchance I speak too boldly, for neither the learned man, nor *Galen* himself, could by any experience ever behold the sensible *Anatomoses*, or ever could demonstrate them to the sense.

I did look after them with all possible diligence, and was not at a little charge and pains in the search of the *Anastomoses*; yet could I never find that any vessel, namely the arteries, together with the veins, were joyn'd by their orifices: I should willingly learn from others who ascribe so much to *Galen*, that they dare swear all which he says. Nor is there any *Anastomosis* in the liver, milt, lungs, reins, or any other of the intrals, although I did boyl them till the whole Parenchyme was made mouldering, and like dust was shaken off, and taken away with the point of a needle, from all the fibers of the vessels, so that I could see the fibers, and the last grains of every division. I dare therefore boldly affirm, that neither the *vena porta* has any *Anastomoses* with the *cava*, nor the veins with the arteries, or the capillar branches of the pore of the choller-bag, which are dispers'd about all the flat of the liver with the veins. Only this you may observe in a fresh liver, that all the branches of the *vena cava* which creep through the whole bunch of the liver, have tunicles pierc'd with many holes, like a sieve, as it is in a sink, fram'd so for the receiving of the blood which falls down. The branches of the Porta are not so, but are divided into stems, and how that both the divisions of these vessels, the one in the flat, the other in the gibbous part, doe run round to the very furthest rising of that intrall without any *Anastomoses*.

Only in three places do I find that which is equivalent to an *Anastomosis*. There rises in the brain, from the soporall arteries creeping down into the Basis, many and unintangled fibers, which afterwards make up the *plexis chorois*, and passing through the ventricles doe at last end in the third receptacle, which performs the office of a vein. In the spermaticall vessels, commonly call'd preparatory, little arteries drawn from the great artery do adhere to the veins preparatory aforesaid, which they accompany, and at last are so receiv'd within their Tunicle, so that at the first they seem both to have one and the same; so that when they end at the upper part of the testicles, where that part passes forth into a point, which is called the varicous and vine-like body, we know not what to call them, veins or arteries, or the ends of both. As likewise the last appearances of the arteries, which go to the Umbilical vein, are obliterated in the Tunicles of that vein.

What doubt is to be made, if through such gulphes, the little branches of the *arteria magna*, swoln with the impulsion and instuffing of blood, could be eas'd of so great and so conspicuous a stream? Nature at least would never have denied us visible and sensible passages, sinks and whirlpools, if she had had intention to have turned all the flux of the blood thither, and by that meanes have deprived the lesser branches, and the solid parts of the benefit of the influx of blood.

Lastly, I will set down one experiment, which seems to be sufficient for the clearing of the *Anastomoses*, and for the overthrowing of their use, and of the passage of the blood, and return of it out of the veins into the arteries by those wayes.

Opening the breast of any creature, and tying the *vena cava* by the heart, so that nothing can passe that way into the heart, and presently cutting the jugular arteries, not touching the veins on neither side, If by giving vent you see the arteries emptied, and not the veins too, I hope it will be clear that the blood is carried out of the veins into the arteries, no where but through the ventricles of the heart: Otherwise (as *Galen* has observ'd) in a little space we should see the veins emptyed, and destitute of blood by the efflux of the arteries.

In what remains, *Riolan*, I both congratulate my self and you; my self for your opinion with which you have adorn'd my Circulation; as likewise I return to you exceeding thanks for your learned, neat, succinct piece which you sent to me, than which there is nothing more elegant; and I both owe and desire to return deserv'd commendation, but I confess, I am not able for such a charge. For I know the name of *Riolan* will afford more praise to me in its subscription, than my prayses, which I wish as great as may be, can do to his *Enchiridion*. The famous book shall outlive all memory, and shall recommend your worth to Posterity, when all Monuments shall perish. To it you have very handsomely adjoyn'd the Anatomy of Diseases, and have very profitably enrich'd it with a new Treatise concerning the Bones. May you, most worthy Man, continually increase in this your worth, and love me, who wish that you may be both happy and long liv'd, and that your most famous writings may be an eternall Commendation to you.

WILLIAM HARVEY.

ANOTHER EXCERCITATION

TO

JOHN RIOLAN.

In which many Objections against the Circulation of the Blood are refuted.

MOst learned *Riolan*, by the help of the Presse, many years ago, I published a part of my labour: But since the birth-day of the Circulation of the Blood, almost no day has past, nor the least space of time, in which I have not heard both good and evill of the Circulation of the Blood which I found out: Others rail at it, as a tender babie unworthy to come to light; Others say, that it's worthy to be foster'd, and favour my writings, and defend them; Some with great disdain oppose them; Some with mighty applause protect them; Others say, that I have abundantly by many experiments, observations, and ocular testimony, confirm'd the Circulation of the blood against all strength and force of arguments; Others think it not yet sufficiently illustrated, and vindicated from objections; But there are who cry out, that I have affected a vain commendation in dissection of living creatures, and do with childish slighting dispraise and deride at Frogs and Serpents, Gnats, and other more inconsiderable creatures brought upon the Stage, and refrain not from ill language. But I think it a thing unworthy of a Philosopher and a searcher of the truth, to return bad words for bad words; and I think I shall doe better and more advised, if with the light of true and evident observations, I shall wipe away those symptomes of incivility.

It cannot be eschewed but doggs will bark and belch up their surfeits; nor can it be help'd but that the Cynicks will be amongst the number of the Philosophers; but we must take a speciall care that they do not bite, nor infect us with their cruel madnesse, or lest they should with their doggs teeth gnaw the very bones or principles of truth.

Detractors, Momes and writers stain'd with railing, as I never intended to read any of them, (from whom nothing of solidity, nor any thing extraordinary is to be hop'd for, but bad words) so did I much less think them worthy of an answer: Let them enjoy their own cursed nature, I beleeve they will find but a few favourable Readers; neither does God give wisdom to the wicked, which is the most excellent gift, and most to be sought for. Let them rail on still, till they be weary (if not asham'd) of it.

If you will enter with *Heraclitus* in *Aristotle* into a work-house (for so I will call it) for inspection of viler creatures, come hither, for the immortal gods are here likewise; and the great and Almighty Father is sometimes most conspicuous in the least and most inconsiderable creatures.

In my book concerning the motion of the heart and blood in creatures, I only chose out those things out of my many other observations, by which I either thought that errours were confuted, or truth was confirm'd; I left out many things as unnecessary and unprofitable, which notwithstanding are discernable by dissection and sense; of which I shall now add some in few words, in favour of those that desire to learn. The great authority of *Galen* is of so much account with every body, that I see many make a difficulty, as concerning that experiment of *Galen* of the ligature of the artery above the pipe, thrust within the concavity of the artery; by which it is demonstrated, that the pulse of the artery comes from the faculty pulsifick, and that it is transmitted from the heart by the tunicles, and not by the impulsion of the blood within the Concavities; and therefore that the arteries are stretch'd as bellows, not as baggs.

This experiment is mentioned by *Vesalius,* a man very skilful in Anatomy; but neither *Galen* nor *Vesalius* says, that they tryed this experiment, which I did; only *Vesalius* prescribes it, and *Galen* counsells it to those that are desirous to find out the truth, not thinking, nor knowing the difficulty of that businesse, nor the vanity of it when it is done, since although it be perform'd with all manner of diligence, it makes nothing to the confirmation of that opinion, which affirms, That the tunicles are the cause of pulsation, but rather shows, That it is set a-work by the impulsion of the blood. For so soon as above the reed, or pipe, You have with a band tied the artery, the artery above the ligature is presently dilated by the impulsion of the blood beyond the mouth of the pipe, from whence both the flux is stop'd and the impulsion reverberated, so that the artery under the band does beat with very little appearance, because the force of the passage of the blood does no way assist it, because it is return'd above the ligature; but if the artery be cut off below the pipe, you shall see the contrary, from the leaping of the blood which is thrown out, and driven through the pipe, as in an Aneurism I have observ'd to come from the exesion of the tunicles of the artery, this (whilst the blood is contain'd within the membranes) hath a contentive vessel of its flux prænaturally made, not of the dilated tunicles of the artery, but of the circumposition of the membrane and flesh. You shall see the inferiour arteries beyond this Aneurism beat very weakly, whilst above, and especially in the Aneurism it self, the pulsations appear great and vehement, although we cannot there imagine, that the impulse or dilatation is made by the tunicles of the artery, or by communication of the faculty of the Cyst, but meerly by the impulsion of the blood.

But that the error of *Vesalius*, and the small experience of others, may the more clearly appear, who affirm (as they imagine) that the part under the pipe does not beat when the band is tyed, I speak by experience, if you make the experiment rightly, that it will; and whereas they say, that upon the untying the band, the arteries below do beat backwards, I say, that the part below beats lesse when you have untyed it, than when it is tyed.

But the effusion of blood which leaps out of the wound confuses all, and makes the experiment vain and to no purpose, so that there can be no certainty demonstrated, as I said, by reason of the blood. But if (and this I know by experience) you lay open the artery, and hold with your finger close that part which you cut, you may at your pleasure try many things which will evidently make the truth appear to you. First, you shall feel the blood, being forc'd, coming down into the artery, by which you shall see the artery dilated; as likewise you may squeez out and let go the blood as you please: If you open a little part of the orifice and look narrowly to it, you shall see the blood at every pulse to be thrown out with a leaping, and as we said in the opening of an artery, or in the perforation of the heart, you shall see the blood to be thrown out in every contraction of the heart, in the dilatation of the artery.

But if you suffer it to flow with a constant and continuall flux, and give it leave to break out, either through the pipe, or by the open orifice, in the streaming of it, both by your sight and by your touch, you shall find all the strokes, order, vehemency, and intermission of the heart; just as you might feel in the pulse of your hand water squirted through a syringe at divers and severall shootings, so you may perceive, both by your sight and by its motion, the blood leaping out with a varying and unequall force. I have seen it sometimes in the cutting of the jugular artery break out with such force, that the blood being forc'd against the hand, did by its reverberation and refraction, flye back four or five foot.

But that this doubt may be more clear, that the pulsifick force does not flow through the Tunicles of the arteries from the heart, I have a little piece of the artery descendant, together with two crural branches of it, about the length of a span, taken out of the body of a very worthy Gentleman, which turn'd to be a bone like a pipe, by the hollow of which, whilst this worthy Gentleman was alive, the blood in its descent to the feet did agitate the arteries by its impulsion; in which case neverthelesse, although the artery were in the same condition, as if it had been bound or tyed above the little conduit-pipe, according to the experiment of *Galen*, that it could not either be dilated in that place, nor streightned like a pair of bellowes, nor from the heart derive its pulsifick force to the inferior or lesser arteries, nor yet carry through the solid substance of the bones that faculty which it had not receiv'd; yet I very well remember that I often observ'd whilst he was alive, that the pulse of the inferiour artery did move in his legs and feet: wherefore it must needs follow, that in that worthy Gentleman, the inferiour arteries were dilated by the impulsion of the blood, like baggs, and not like bellows, by the stretching of the tunicles. For there must needs arrive the same inconvenience, and interception of the pulsifick faculty, the tunicle of the artery being wholly converted into a conduit, or pipe of bone, as might arrive from the reed or pipe which was tyed, that the artery might not beat.

I knew likewise in another worthy and gallant Gentleman, the *aorta*, and a part of the great artery near the heart, turn'd into a round bone. So *Galen's* experiment, or at least one answerable to it, being not found out by industry, was found out by chance, and does manifestly evidence, that the interception of the pulsifick faculty is not intercepted by the constriction or ligature of the tunicles, so that by that means the arteries cannot beat; and if the experiment which *Galen* prescribes, were rightly perform'd by any, it would refute the opinion which *Vesalius* thought from thence to have confirm'd. Yet for this cause we do not deny all motion to the tunicles of the arteries, but do attribute that to it, which we grant to the heart, namely, that there is a coarctation, and a Systole in the tunicles themselves, and from their distension a regress to their natural constitution. But if this is to be observ'd, that they are not dilated and streightned for the same cause, nor by the same instrument, but by severall, as you may observe in the motion of all the parts, and in the heart; it is distended by the ear, contracted by it self, so the arteries are dilated by the heart, and fall of themselves.

So you may make another experiment after the same manner: If you fill two sawcers of the same measure, one of them with arterial blood, which leaps out, the other with venal blood, drawn out of a vein of the same Animal, you may presently by your sense, and afterwards too, when both the bloods are grown cold, observe what is the difference betwixt both the bloods, against those who do fancy another sort of blood, in the arteries than is in the veins; namely, they do ascribe to the veins a fresher sort of blood, I do not know which way boyling or blown up, swelling or bubbling (like to honey or milk upon the fire), and so taking up more room.

For if the blood which is driven out of the left ventricle into the arteries should be leaven'd, so as to be blown up, and foam after that manner, so that a drop or two should fill all the concavity of the aorta, no doubt it would when it fell again, return to the quantity of some few drops (which cause some do allege for the emptiness of the arteries in dead men) and the same would be seen in the *cotyla* full of arterial blood; for so we find that it comes to passe in the cooling of milk or honey. But if in either *cotyla* the blood be found of the same colour, and congealed, of a not much different consistence, and squeezing out the whey after the same manner, and if it take up the same room both when it is hot and when it is cold, I think it will be a sufficient argument to gain any man's belief, and to confute the dreams of some, that there is neither in the left ventricle any sort of blood differing from that of the right (as you may find out both by sense and reason), for you must needs likewise affirm, that the *vena arteriosa* should equally be distended with one drop of blood foaming up, and therefore that there is just such bubbling and leaven'd blood in the right as in the left, seeing the entry of the *vena arteriosa*, and the egress of the aorta, is equipollent and equall.

Three things are chiefly ready to breed this opinion of the diversity of blood. One is, that in the cutting of an artery, they see brighter blood drawn out: Another is, that in the dissection of dead bodies, they find both the left ventricle of the heart, and all the arteries so empty: A third is, that they imagine that the arterial blood is more spirituous, and more replete with Spirits; and therefore they think that it takes up more room: The cause and reasons of all which things why they come to be so, by inspection is perceiv'd.

First, insomuch as concerns the colour, always and every where blood coming through a narrow hole, is as it were strained and becomes thinner, and the lighter part of it, and which swims above, and is more penetrable, is thrust out: so in Phlebotomy, the blood which springs out with greater flux or force, and out of a greater orifice, and flies further, is always thicker, fuller, and darker colour'd; but

if it flow out of a little and narrow hole, and by
drops (as it does out of a vein, when the ligature
is unty'd), it is brighter, for it is strain'd as it were,
and only the thinner part comes out, as in the
bleeding at nose, or that wich is extracted by Leeches,
or Cupping-glasses, or any way issuing by *diapedesis*,
is always seen more bright; because the thicknesse
and hardnesse of the tunicles becomes more impas-
sible, nor yeelds so pliably, as to give an open way
for the coming out of the blood: as it likewise
happens in fat bodies, when by the fat under the skin
the orifice of the vein is stop'd, then the blood
appears thinner, brighter, and as if it did flow from
an artery. On the contrary, if you receive in a
sawcer the blood, when you have cut an artery, if
it flow freely, it shall appear like venal blood; there
is blood much brighter in the lungs, and squeez'd
out from thence, than any is found in the arteries.

The emptiness of the arteries in dead bodies
(which did perchance cozen *Erasistratus*, insomuch
that he thought that the arteries contain'd only aerial
spirits) proceeds from hence, because that when the
lungs fall (their passages being stopt) the lungs do
breathe no longer, so that the blood cannot freely
pass through them, yet the heart continues a while
in its expulsion, whence both the left ventricle of
the heart is more contracted, and the arteries likewise empty, and not fill'd by succession of blood,
appear empty; But if the heart cease both at one
time, and the lungs to give passage by respiration,

as it is in those who are drowned in cold water, or in those who are taken suddenly with unexpected death, you shall find both the veins and the arteries full.

As concerning the third, of the Spirits, what they are, and of what consistence, and how they are in the body, whether they be apart and distinct from the solid parts, or mix'd with them, there are so many and so diverse opinions, that it is no wonder if Spirits, whose nature is left so doubtfull, do serve for a common escape to ignorance: For commonly ignorant persons when they cannot give a reason for any thing, they say presently, that it is done by Spirits, and bring in Spirits as performers in all cases; and like as bad Poets do bring in the gods upon the Scene by head and ears, to make the *Exit* and *Catastrophe* of their play.

Fernelius and others do imagine aerial Spirits, and invisible substances; for he proves that there are animal Spirits (just as *Erasistratus* proves them in the arteries) because there are little cells in the brains which are empty, and since there is no *vacuum*, he concludes, that in living men they are full of Spirits.

Yet all the School of Physicians agrees upon three sorts of Spirits, that the natural Spirits flow through the veins, the vital through the arteries, and the animal through the nerves; whence the Physicians say out of *Galen*, that the parts sometimes want the consent of the brain, because the faculty, together with its essence, is sometimes hinder'd, and sometimes without the essence. Over and above besides these three sorts of influxive spirits, they seem to assert so many more, which are implanted. But none of all these have we found by disseǎion, neither in the veins, nerves, arteries, nor parts of living persons. Some make corporeal Spirits, others some incorporeal Spirits; and those who make corporeal Spirits, sometimes say, that the blood, or thinnest part of the blood, is the conjunǎion of the soul with the body; sometimes they say, that the Spirits are contained in the blood (as flame in smoke) and sustain'd by the perpetuall flux of it; sometimes they do distinguish them from the blood. Those that affirm that there are Spirits incorporeal know not how to tread, but likewise doe affirm that there are potential Spirits, as Spirits concoǎive, chilificative, procreative, and so many Spirits as there are faculties or parts.

But the Schoolmen tell us also of a Spirit of Fortitude, Prudence, Patience, and of all the vertues, and the most holy Spirit of wisdom, and all divine gifts. They think too that bad and good Spirits do assist, possess, leave, and wander abroad. They think also, that diseases are caus'd by a Devil, as by a *Cacochima*. But although there is nothing more uncertain and doubtfull than the doǎrine which is assign'd to us concerning the spirits: yet for the most part all Physicians seem with *Hippocrates* to conclude, that our bodies are made up of three parts, containing, contain'd, and enforcing; by the forcing he means Spirits.

But if Spirits must be understood to be every thing which enforces in a man's body, whatsoever hath the power or force of action in living bodies must be call'd by the name of Spirits. Therefore all the Spirits are not aeriall substances, nor powers, nor habits, nor incorporeal.

But omitting the tediousnesse of all other significations to our purpose, those Spirits which passe out through the veins or the arteries, are not separable from the blood, no more than flame from the flakes about it. But the blood and the Spirit signifie one and the same thing, though divers in essence, as good Wine and its Spirit. For as Wine is no more Wine after it has lost its Spirit, but flat stuff or vinegar, so neither blood without Spirit is blood, but equivocally goar; as a hand of stone or a dead hand is no more a hand, so blood without vital spirit is no more to be esteemed blood. So the Spirit which is chiefly in the arteries, and the arterial blood is as its act, as the Spirit of Wine in Wine, and the Spirit of *Aqua vitæ*, or as a little flame kindled in the Spirit of Wine, and living by nourishing of it self.

Therefore blood when it is most imbued with Spirits, it does require and look after more room, because it is swell'd or leaven'd, and blown up by them (which you may certainly judge in my experiment which I brought concerning the measure of the sawcers); but like wine, because it has greater strength and force of action and performance, in which it excels, according to the mind of *Hippocrates*.

Therefore the same blood is in the veins which is in the arteries, though it be acknowledg'd to be more full of Spirit, and more eminent in vital force: but it is not converted into something more aerial or vaporous, as if there were no Spirits but aerial ones, or none that had force but such as were flatuous and windy: But neither are the Animal Spirits natural and vital, which are contained in the solid parts, to wit, the ligaments and nerves (especially if there be so many severall sorts of them), thought to be so many aerial forms, or divers sorts of vapours.

Those who acknowledge Spirits in the bodies of creatures, but such as are corporal, but of an aerial consistence, or vaporous or fierie, of them would I fain know, Whether they can passe hither and thither, backward and forward, as distinct bodies without the blood? Whether or no, I say, the Spirits follow the motion of the blood, as if they were either parts of the blood, or adhering to it by an indissoluble connexion, and an interrupted exhalation; so that they can neither leave the parts, nor passe without the influx, reflux, and passing of the blood.

For if, as the vapours attenuated by the heat of the water, the Spirits, by the continuall flux and succession of the blood, become the nourishment of the parts, it will necessarily follow, that they cannot remain apart from the nourishment, but do continually vanish, for that same reason that they neither flow back nor pass any way, nor abide, but according to the influxion, refluxion, or passing of the blood, as being either their subject, *vehiculum*, or nourishment.

Then I would know, how they show us that Spirits are made in the heart, and do make them up, either by the compounding of exhalations, or vapours of the blood (rais'd either by the heat or concussion of the heart). Are not such Spirits to be thought much colder than the blood, since both the parts of which they are compounded, to wit, air and vapour, are much colder? for the vapour of boyling water is much more tolerable than the water it self, and any flame burns less than the coal of a candle, and a wood-coal less then iron or brasse red hot.

Whence it seems that such Spirits doe owe their heat to the blood, rather than the blood is heated by the Spirits, and such Spirits are rather to be deem'd fumes and excrements, flowing from the blood and body (like smels), than workers in Nature; especially since they being so frail and vanishing, do so quickly lose that vertue, which in their original they receive from the blood.

From whence it were likewise probable that there should be an expiration of the lungs, by which these Spirits being blown out might be ayr'd and purified, and that there should be an inspiration into them, that the blood passing through betwixt the two ventricles of the heart might be temper'd by the ambient cold, lest being heated, and rising and swelling with a kind of fermentation, like boyling honey or milk, it should so distend the lungs as to suffocate the creature, as in a dangerous Asthma we have often seen: To which *Galen* likewise ascribes the reason, when he says, that this comes to passe by obstruction of the little arteries, namely the venous and arterious vessels. I have had experience of this, that by affixing of Cupping-glasses, and pouring upon them good store of cold water, there has many been sav'd who have been in danger to be suffocated by an Asthma. I have here, perchance, spoken sufficiently concerning Spirits, which we ought to define, and show what and how they are in a Treatise of Physiologie: so much I will adjoyn.

Those that speak concerning innate warmth, as an ordinary instrument of Nature in performance of all things, and tell us of the necessity of influxive heat, to entertain all the parts, and keep them in life, and do acknowledge that it cannot exist without a subject, because they find a movable bodie disproportionable, by reason of the swiftness of the flux and reflux (especially in the passions of the mind), and because of the swift motion of this heat, they introduce Spirits, as bodies most subtle, penetrative and movable; and just as they say, that from that ordinary instrument, to wit, the innate heat, proceeds the admirable divinity of Natural operations: so doe they likewise affirm, that those Spirits of a sublime, bright, æthereal and celestial nature are the bonds of the Soul; as the ignorant common-people when they do not conceive the reasons of things, think and say that God is the immediate author of them.

Whence they resolve, that the influxive heat does come swiftly through all the parts, by the influx of Spirit, and that it comes through the arteries; as if the blood could not be so speedily mov'd, nor so fully nourish; and in the confidence of this opinion they are so far advanced, that they deny that there is any blood contained in the arteries.

And with this very slight argument they endeavour to ground this, that the arterial blood differs from the blood of the veins, or that the arteries are fill'd with such Spirits and not with blood, contrary to all that which *Galen* both from reason and experience brought against *Erasistratus*.

But it is manifest by our former experiment, and by sense that the arterial blood is not so different, the influx of the blood and Spirit with it being not separate from the blood; but that it flows in one body through the arteries, sense may likewise make evident.

You may observe when, and as often as the extremities of the hands, the feet and the ears are stiff and cold, and are restor'd again by the influx of heat, that it happens that at the self-same time they are colour'd, warm'd and fill'd, and that the veins which were unseen before, do swell to plain appearance, from whence sometimes, when they are suddenly warm'd again, the parts are sensible of some pain; from which it appears, that the same which by its influx brings heat, the same is it that fills and colours them, but this can be nothing else but blood, as was demonstrated before.

Cutting off a long arterie or vein any body may see this evidently by sense, when he shall see the nearer part of the vein towards the heart let out no blood, but the further part pour it abundantly, and nothing but blood (as afterwards in my experiment which I set down, which I tryed in the inner jugularie veins). On the other side, cutting an arterie, but a little blood flows from the further part, but the nearer part shoots with a violent force mere blood, as if it were out of a spout.

By which experiment it is known which way the passage is in them, either this way or that way. Besides, you'l know what swiftnesse there is in it, what sensible motion, not by little and by drops, and with what violence to boot.

But lest any would make an evasion, by pretending of invisible Spirits; Let the orifice of the vessel so dissected be let down into a vessel of water or oyl, for if any aerial thing came out, it would break out by visible bubbles; for after this manner Wasps, Hornets, and the like Insects, being drowned or suffocate in oyl, send out at last bubbles from their tail when they are dying: from whence it is not improbable that they do take breath too whilst they are alive.

For all creatures at last when they are drown'd and stifled in the water, when they fail and sink, they use to send out bubbles out of their mouth and lungs, when they give up the ghost.

Lastly, it is assur'd by the same experiment, That

the portals in the veins are so exactly shut, that air when it is blown in cannot pass, much lesse blood. I say that it appears to the sense, that neither sensibly not insensibly, neither by little nor by drops, the blood is remov'd from the heart by the veins.

And lest any should flye hither, and say thus, That this comes to pass when Nature is troubled, and does act besides Nature, not when she is left to her self, and acts at her own freedom; seeing the same things appeare in a sickly and preternatural constitution, which appear in good estate of bodie, it is not to be said, that cutting off a vein, since there flows so much blood from the further part, that this comes to pass beside Nature, because Nature is molested; for the dissection does not shut the further part, so that nothing can get out that way, nor can it be squeez'd out whether Nature be troubled or no. Others doe wrangle after the same manner, saying, That although when the arterie is cut near the heart the blood breaks out in so great abundance immediately, yet for that cause the heart being whole, and the arterie too, it does not alwayes drive the blood by impulsion. Yet it is more likely, that all impulsion does drive something, nor can there be a pulse of the container without the impulsion of something contained: Yet some, that they might defend themselves, and decline the Circulation of the blood, are not afraid to affirm and maintain this; to wit, that the arteries in living creatures, and being according to Nature, are so full that they cannot receive a grain weight more of blood: and so likewise of the ventricles of the heart. But it is without doubt, whensoever, or how much soever the arteries and ventricles are dilated, and contracted, they ought to receive greater impulsion of blood, and that beyond many grains. For if the ventricles be so distended as we have seen in the Anatomie of living creatures till they receive no more blood, the heart leaves beating, and continuing stiff and resisting, it occasions death by suffocation.

Whether the blood be mov'd or driven, or move it self by its own intrinsecall nature, we have spoken sufficiently in our book of the motion of the heart and blood; as also concerning the action, function, contraction, dilatation of the heart, how it is done, and together with the Diastole of the arteries, so that those which take arguments from thence for contradiction, seem either not to understand what is said there, or else they will not try the business by their own sight.

I believe there can not the attraction of any thing be demonstrated in the body but of the nutriment, which by succession of parts supplies by little and little that which is lost, as the oyl of a lamp by the flame.

Whence that is the first common organ of all sensible attraction and impulsion, which has the nature of a nerve, or of a fiber, or of a muscle, to wit, that it may be contracted and that by shortning of it self it may stretch out, draw in, or thrust forward: but these things are more fully and openly to be declared elsewhere, in the organs of motion in living creatures.

Insomuch as to those who do still reject the Circulation, because they neither see the efficient, nor finall cause of it, there remains, because I have as yet joyn'd nothing to it, only to say thus much: First you must confess that there is a Circulation, before you enquire for what it is, for from those things that doe happen upon the circulation and allowance of it, the use and profits accrewing are to be searched. In the mean time I shall say so much, that there are many things allowed and received in Physiologie, Pathologie, and Medicine, that no body knows the cause of; yet that there are such things no body is ignorant, namely, of rotten feavers, revulsion, purgation of excrement, yet all these things are known by the help of Circulation.

Whosoever therefore does oppose the Circulation of the blood, beeause so long as the Circulation stands, they cannot resolve Physicall Problems, or because in curing of diseases, and using of medicaments, they cannot from thence assign any cause of the Symptomes, or see that those causes which from their Masters they have receiv'd, are false, or think it an unworthy thing to desert opinions approved heretofore, and think it unlawful to call in question the discipline which has been receiv'd through so many ages, together with the authority of the Antients.

To all these I answer that the deeds of nature,

which are manifest to the sense, care not for any opinion or any antiquity, for there is nothing more antient then nature, or of greater authority.

Besides, those Problemes out of Medicinall observations not to be solv'd, as they Imagine, to the Circulation they object, and do oppose to it the declaring of their own errours, to wit, that if the circulation be true there can be no revulsion, since the blood is driven upon the part affected as before, and so it is to be feared, that there will be a passage of the excrements and blood, through the most noble and principall of our entrails. They do admire at the efflux and excretion, when out of the same body at divers holes, yea sometimes at the same hole, foul and corrupt blood issues, whereas if the blood were driven with a continuall flux, passing through the heart, it would be mix'd and shaken together.

They do doubt how these, and many other things that they fetch from the School of Physicians can come to pass, for they seem to be repugnant to the Circulation of the blood, nor do they think (as it is in Astronomie) that it is enough to make new Systemes, unless you solve all scruples.

I thought fit to return no other answer at this time, but that the Circulation is not the same every where, and at all times, but many things do happen from the swifter or slower motion of the blood, either through the strength or infirmity of the heart which drives it, by the abundance, estate or constitution of the blood, the thickness of the parts, obstruction, and the like; thicker blood hardly finds way through narrow passages; it is more strained when it passes the streyner of the liver, than when it passes the streyner of the lungs.

It does not with a like speed passe through the thin contexture of the flesh and parenchyme, as it does through the thick consistence of the nervous parts. For the thinner, more pure, and more spiritous part is sooner streyn'd through; the more earthy, cacochymick, and more tardy, stayes longer and is turn'd back. The nutritive part and last aliment (be it the *Ros* or *Cambium*) is more penetrative, seeing it is to be applied to every part, whether it be to the horns, feathers, or nayls, if being every where nourished they increase in all their dimensions; for this reason the excrements in some places are voyded, thickned, or do burthen us, or are concocted: Nor do I think that there is any necessity that the excrements, or ill humours, being once set apart, nor the milk, flegm, nor sperm, or the last nutriment (the *Ros* and *Cambium*) should be return'd with the blood, but that it behoves that that which nourishes should adhere, that it may be agglutinated. Of which, and a great many other things which are to be determined and declar'd in their proper places, to wit, in Physiologie, and the rest of the parts of Physick, it is not fit to dispute, nor yet of the consequences of the Circulation of the blood, nor the conveniences nor inconveniences of it, before the Circulation it self be established for granted.

The example of Astronomie is not here to be followed, where only from appearances, and such a thing that may be, the causes, and why such a thing should be, comes to be enquir'd after. But as one desiring to know the cause of the Eclipse, ought to be plac'd above the Moon, that by his sense he might find out the cause, not by reasoning of things sensible, in things which come under the notion of sense, no surer demonstration can be to gain beleef, than ocular testimony.

I desire that there may be one other remarkable experiment tryed by all that are desirous of the knowledge of the truth, by which likewise the pulse of the arteries is both seen to be done by the blood, and evidenced to be so.

If the Gutts of a dog, or a wolf, or any Creature stuff'd, and dryed, such as you see at the Apothecaries, you cut away a part of it of any length, and fill it with water, and tie it at both ends, that it is like a pudding, hitting or shaking the one end of it, in the end over against it, by putting two of your fingers (as we use to feel the pulse of the artery above the wrist) you may find every stroke and difference of the motion clearly. And after this manner in every swelling vein, either of living or dead, you may to raw students manifest all the differences of the pulses to the sense, in greatness, frequency, vehemency, and time. For as it is in a long bladder, or in a long drum, all the strokes of one of the extremes is felt likewise in the other;

Therefore in the Hydropsie of the belly, as likewise in all abscessions which are fill'd with liquid
matter, we use to distinguish an *Anasarca* from a
Tympanitis; If all pulses and vibrations made in one
side, be by touch clearly felt in the other, we think
it a *Tympanitis*, and not as it is falsely beleev'd because
it is like the sound of a drum, and is only by flat-
uousness, but because (as it is in a drum) every
light stroke passes through it, and every shake goes
through the whole; for it shews that there is a serous
and wheyish substance within, and not a tough and
slimy, as in the *Anasarca*, which being thrust retains
the marks of the stroke or impulsion, and transmits
it not. Having opened this experiment, there rises
a most powerfull objection against the Circulation
of the blood, neither observ'd, nor oppos'd against
me by any that has hitherto written. Seeing in this
experiment we see that there may be Systoles and
Diastoles, without the egresse of the liquor, who
will beleeve but that it may be just so in the arteries
and that in them just so as it is in an *Euripus*, from
hence thither, and from thence hither, it may be
driven by turns. But in another place we have sufficiently resolv'd this doubt, and now we also say,
that this is not so in the arteries of living creatures,
because continually and incessantly the right ear of
the heart fills the ventricles with blood, the return
of which the three-pointed portals hinder, and so the
left ear fills the left ventricle, and both the ventricles in the Systole throw forth the blood which the

Sigmoidal portals hinder to return, and that it ought therefore either passe some way, and continually out of the lungs and arteries, or otherwise it would at last, by restagnation and intrusion, break the vessels which contain it, or suffocate the heart it self by distention, as we have observ'd to be plain to the sense in the dissection of a live Adder, in my Book concerning the motion of the blood.

To clear this doubt, I will recite to you two experiments, amongst many other (of which I told one before) by which it clearly appears, that the blood in the veins with a continuall and great flux runs continually towards the heart.

In the internal jugular vein of a live Doe, which I laid open before a great part of the Nobility, and the King my Royal Master standing by, which was cut and broke off in the middle: From the lower part rising from the Clavicule, scarce a few drops did issue, whilst in the mean time the blood with great force, and breaking out of a round stream, ran out most plentifully downards from the head through the other orifice of the vein. You may observe the same daily in Phlebotomy in the flowing out of the blood, if you hold the vein fast with one finger a little below the orifice, presently the flux is stopped, which after you let it go flows abundantly, as before.

In any visible long vein of your arm, stretching out your hand, and pressing out all the blood downwards as much as you can, you shall see the vein fall, leaving as it were a furrow in the place, but so soon as you thrust it back with one of your fingers, you shall presently see the part towards the hand, to be fill'd and swell, and to rise by the return of the blood from the hand. What is the reason, that by stopping of the breath, and by that means streightning the lungs, and a great deal of air being within, the pectoral vessels are streightned, whence the blood is driven into the face and eyes with so much rednesse?

Nay that (as *Aristotle* says in his Problemes) all actions are perform'd with greater strength by keeping in of the breath, than by letting it free ; so you get blood more abundantly out of the veins of the brow, or tongue, by compression of the throat and retention of breath.

I have found sometimes in a man's body, newly hang'd, two hours after his execution, before the rednesse of his face was gone, opening up his heart and Pericardium, the right ear of his heart and lungs much stuffed, and distended with blood; many witnesses standing by, especially I shew'd them the ear, as big as a man's fist, so swel'd, that you would have thought it would have burst with greatnesse which, the body being afterwards cold, and the blood having found other ways, was quite gone.

So from these and other experiments, it is clear enough, that the blood runs through all the veins to the basis of the heart, and that unless it found passage, it behov'd to be streightned, or shut up in other ways, and that the heart would be o'erwhelmed with it, as on the other part, if it did not flow out of the arteries, but were regurgitated, the oppression by it would quickly appear.

I will add another observation: A noble Knight Baronet, Sir *Robert Darcie*, father to the Son-in-Law of the most learned man, and my very great friend, and a famous Physician, Dr. *Argent*, about the middle of his age, did often complain of an oppressive pain in his breast, especially in the night time, so that sometimes being afraid of collapsion of spirits, sometimes fearing suffocation by a Paroxisme, he led an unquiet and anxious life, using the Counsell of all Physicians, and taking many things in vain; at last the disease prevailing, he becomes cacheſtick, and Hydropick, and at last opprest in a signal Paroxism, he dyed. In his Corps, in the presence of Dr. *Argent*, who at that time was President of the College of Physicians, and Dr. *Gorge*, a rare Divine, and a good Preacher, who was at that time Minister of that Parish, by the hinderance of the passage of the blood out of the left ventricle into the arteries, the wall of the left ventricle it self (which is seen to be thick and strong enough) was broken, and poured forth blood at a wide hole, for it was a hole so big, that it would easily receive one of my fingers.

I knew another stout man, who did so boyl with rage because he had suffer'd an injury, and receiv'd an affront by one that was more powerfull than himself, that his anger and hatred being increas'd every day (by reason he could not be reveng'd) and discovering the passion of his mind to no body, which was so exulcerate within him, at last he fell into a strange sort of a disease, and was tortur'd, and miserably tormented with great oppression and pain in his heart, and breast, so that the most skilfull Physicians' prescriptions doing no good upon him, at last, after some years, he fell sick of the Scorbutick disease, pin'd away, and dyed.

This man only found ease as oft as his brest was prest down by a strong man, and was thump'd and beaten down as they do when they mould bread: his friends thought he was bewitched, or possessed with the Devil.

He likewise had his jugular arteries distended about the greatnesse of one's thumbs, as if either of them had been the *Aorta* it self or the *Arteria magna* in its descent, and did beat vehemently, and were to the view like two long Aneurisms, which caus'd us try blood-letting in his temples, but that gave him no ease. In his corps I found the heart and the *aorta* so distended and full of blood, that the bigness of his heart, and the concavities of the ventricles, were equall in bigness to that of an Oxe; so great is the strength of the blood when it is shut up, and so vast its force.

Although then (by the experiment newly mention'd) there may be an impulsion without an exite (in the shaking of water up and down) in the pudding afore mentioned, yet cannot it be so in the blood which is in the vessels of living persons, without very great and heavy impediments and dangers.

Yet from thence it is manifest that the blood in its Circulation does not passe everywhere with the same agility and swiftnesse, nor with the same vehemence in all places and parts, and at all times, but that it varies much according to the age, sex, temper, habit of body, and other contingents, external, internal, natural, or preternatural.

For it does not pass through the crooked and obstructed passages, with the same swiftnesse as it does through those that are open, free, and patent; nor does it passe through bodies or dense parts, and such as are stuff'd or constricted, as it does through those that are thin, open, and without obstruction; nor does it run out so swiftly and penetratively when the impulsion is slow and soft, as when it is driven with force and strength, and thrust forward with vehemency and abundance. Nor is the thick blood or solid masse, or when it is made earthy, so penetrative, as when it is more wheyish, made thin and liquid.

And therefore with reason we may imagine, that the blood in its Circulation goes slowlier through the reins, than through the substance of the heart; swiftlier through the liver, than through the reins; swiftlier through the spleen, than through the liver; swiftlier through the lungs than through the flesh, or any other viscers of thinner contexture.

We may likewise contemplate in the age, sex,

temperature, habit of the body, soft or hard, of the ambient cold, which condenses bodies, when the veins scarce appear in the members, or the sanguine colour is seen, or the heat appears, the blood being made more liquid by reception of nutriment. So likewise the veins do more conspicuously, and freely pour out the blood, the body being heated before opening of a vein, than when it is cold. We see that the passion of the mind (in the administration of Phlebotomie) if any fearful person chance to sound, streight the flux of the blood is stopp'd, and a bloodless paleness seizes on all the superfice of his body, his members are stiff, his ears sing, his eyes grow dim, and are in convulsion. I find here a field where I might run out further, and expatiate at large in speculation: But from hence so great a light of truth appears, from which so many questions may be resolv'd, so many doubts answered, so many causes and cures of diseases found out, that they seem to require a particular treatise. Concerning all which in my medicinal observations, I'll set down things worthy your admiration.

For what is more admirable, than that in all affections, desires, hope, or fear, our bodies suffer severall ways, our very countenances are changed, and our blood is seen to fly up and down? with anger our eyes are red, the black of the eye is lessen'd in shamefastnesse, and the cheeks are flush'd with redness; by fear, infamie, and shame, the face is pale, the ears glow, as if they should hear some ill thing:

Young men that are touched with lust, how quickly is their nerve fill'd with blood, erected and extended? But it is most worthy the observation of Physicians, why bloodletting and cupping glasses, and the stopping of the arterie which carries the flux (especially whilst they are doing) does as it were with a charm take away all pain and grief: I say, such things as these are to be referred to observations, where they are explained clearly.

Frivolous and unexperienced persons do scurvily strive to overthrow by logicall and far-fetch'd arguments, or to establish such things as are meerly to be confirm'd by Anatomical dissection, and ocular testimony. It behoves him, who ever is desirous to learn, to see any thing which is in question, if it be obvious to sense and sight, whether it be so or no, or else be bound to believe those that have made tryall, for by no other clearer or more evident certainty can he learn or be taught. Who will perswade a man that has not tasted them, that sweet or new wine is better than water? with what arguments shall one perswade a blind man thar the Sun is clear, and out-shines all the Stars in the firmament? So concerning the Circulation of the blood, which all have had confirm'd to them for so many years, by so many ocular experiments, there has been hitherto no man found, who by his observations could refute a thing so obvious to the sense (to wit the motion of flux and reflux) by observations alike obvious to the sense, or destroy the confirm'd experience of it, nay by ocular testimony none ever offer'd to build up a contrary opinion.

Whilst in the mean time there are not wanting persons, who for their unskilfulnesse and little experience in Anatomie, having nothing agreeable to sense to oppose to it, they cavill at it with some vain assertions, and such as they adhere to from the authority of Teachers, with no solid supposition, but with idle and frivolous arguments, and bark at it besides with a great many other words, and those base ones too, with rayling, and base scurvy language, by which they do no more than shew their own vanity and folly, and their baseness and want of arguments, which are to be fetch'd from sense, so that they with their false Sophisticall arguments do rage against sense: Just as when the raging winds advancing the waves in the Sicilian Sea dashes them in pieces against the rocks within *Charybdis*, they make a hideous noise, and being broken and reverberated, hisse and foam; so do these men rage against the reason of their own sense.

If nothing should be admitted by sense without the testimony of reason, or sometimes against the dictate of reason, there should be no question now to be controverted.

If our most certain Authors were not our senses, and these things were to be established by reasoning, as the Geometricians do in their frames, we should truly admit of no Science, for it is the rationall demonstration of Geometrie from things sensible to demonstrate things to the sense, according to which example, things abstruse, and hid from the sense, grow more manifest by things which are easier, and better known. *Aristotle* advises us much better, *lib.* 31, *de Generatione Animalium*, disputing of the generation of Bees, says he, *you must give credit to your senses: if those things which are demonstrated to you are agreeable to those things which are perceptible by sense, which, as they shall then be better known, so you may better trust your sense than your reason.* Whence we ought to approve or reject all things by examination leisurely made, but if you will examine or try whether they be said right or wrong, you must bring them to the test of sense, and confirm, and establish them by the judgement of sense, where, if there be any thing feign'd or not, sure it will appear. Whence *Plato* says in his *Critias*, That the explication of those things is not hard, of which we can come to the experiment, nor are those auditors fit for Science that have no experience.

How hard and difficult a thing is it for those that have no experience to teach such things of which they have no experience, or sensible knowledge; and how unfit and indocile unexperienced Auditors are to true Science, the judgement of blind-men in colours, and of deaf men in the distinction of sounds, does plainly shew. Who shall ever teach the flux and reflux of the Sea? or by a Geometrical Diagram teach the quantities of Angles, or the computation of the sides of a figure to a blind-man, or to those that never saw the Sea, nor a Diagram? A man that is not expert in Anatomie, in so far as he cannot conceive the businesse with his own eyes, and proper reach, in so far is thought to be blind to learning, and unfit; for he knows not truly any thing concerning which an Anatomist disputes, nor any thing, from the implanted nature of which he should take his argument, but all things he is alike ignorant of, as well those things which are gathered and concluded, as the things from whence. But there is no possible knowledge, which arrives not from a pre-existent knowledge, and that very demonstrable. This one cause is the chief reason why the knowledge we have of the heavenly bodies is so uncertain and conjectural. Very fain would I know from those ignorant persons, that professe the causes and reasons of all things, why as both the eyes in beholding move together every way, nor particularly one moves this way, and the other that way, so neither both the ears of the heart?

Because they know not the causes of fevers, or of the plague, or the admirable properties of some medicaments, and the causes why they are so, must therefore these things be denyed?

Why is the Birth that breathes not till the tenth moneth not suffocated for want of ayr? since one that is born in the seventh or eighth, so soon as he has breathed in the air, is presently choak'd if it have no air? How can it retain life whilst it is yet within the Secundine, or as yet not come forth, without breath? but so soon as he comes into the air unlesse he breath he cannot live?

Because I see many men doubtful in the Circulation, and some men oppose such things which understand them not aright, as I intended them, I shall briefly rehearse out of my Book of the motion of the heart and blood, what I did there intend. The blood which is contain'd in the veins (as in its own hold) where it is most abundant (to wit, in the *vena cava*) near to the Basis of the heart and the right ear, growing hot by little and little by its own internal heat, and made thin, it swells and rises (like leaven) whence the ear being first dilated, and afterwards contracting it self by its pulsifick faculty, streightways drives it out into the right ventricle of the heart, which being fill'd in its Systole, and consequently freeing it self from that blood which is driven into it (the three-pointed portals refusing passage to it) it drives the same blood into the *vena arteriosa* (where the passage is open) by which it does distend it. Now the blood in the arterious vessel being not able to return against the Sigmoidal portals, but because the lungs are extended, amplified, and restricted both by inspiration and expiration, and likewise their vessels, they give passage to this blood into the *arteria venosa*: of which the left ear keeping together equal motion, time and order, with the right ear, and performing its function, sends the same blood into the left ventricle, as the right sent into the right, whence the left ventricle together, and at the same time with the right (since it can gain no regress, by reason of the portals which hinder its return) drives it into the capaciousnesse of the *aorta*, and consequently into all the branches of the arterie; the arteries being filled with this sudden pulse, being not able so suddenly to disburthen themselves, are distended, suffer an impulsion and Diastole.

Whence I gather, seeing the same is reiterated continually and incessantly, that the arteries, both in the lungs and in the whole body, by so many stroaks, and impulsions of the heart, would be so distended and stuffed with blood, at last, that either the impulsion would give over all together, or else the arteries would burst, or be so dilated, that they would contain the whole mass of blood which is in the veins, unlesse the efflux of blood were disburthen'd somewhere.

We may likewise reason after the same manner of the ventricles of the heart, being fill'd and stuff'd with blood, unlesse the arteries did likewise disburthen they would be at last distended and destitute of all motion. This consequence of mine is demonstrative and true, and followes of necessity, if the premises be true; but our senses ought to assure us whether such things be false or true and not our reason, ocular testimony and no contemplation.

I affirm likewise of the blood in the veins, that the blood does always, and every where, run out of the lesse into the greater, and hastens towards the heart from every part: whence I gather that whatsoever quantity which is continually sent in, the arteries do receive by the veins, that the same does return and does at last flow back thither from whence it is first driven, and that by this means the blood moves circularly, being driven in its flux and reflux by the heart, by whose force it is driven into all the fibres of the arteries, and that it does afterwards successively by a continuall flux return through the veins, from all those parts which draw, and streyn it through; sense it self teaches us that this is true, and collections from things obvious to sense takes away all occasion of doubt.

Lastly, this is that I did endeavour to relate and lay open by my observations and experiments, and not to demonstrate by causes and probable principles, but to confirm it by sense and experience, as by a powerfull authority, according to the rule of Anatomists.

From these we may observe what force, and violence and strong vehemency we perceive in the heart and greater arteries by touch and sight; I do not say, that in all the vessels which contain the blood, the pulse of the Systole and Diastole is the same (in greater Creatures), nor in all creatures which have blood, but that there is such a one and so great in all, that by that means there is a flux of blood, and swifter course of it through the small arteries, the porosities of the parts and branches of all the veins, and from thence comes the Circulation: for neither the small arteries, nor the veins do beat, but only the arteries which are nighest to the heart, because they do not so soon send the blood out as it is driven into them, for you may try, opening of an arterie, if the blood leap out in full stream, so that it come out as freely as it went in, that you scarce found any pulse in that arterie through which it passes, because the blood running through and finding passage, does not distend it. In Fishes, Serpents, and colder creatures, the heart beats slowly and weaker, that you will hardly perceive any pulse in the arteries, because they passe their blood through very slowlie; whence it is that in these as also in the little fibers of the arteries of a man there is no distinction by blood; because they are not pierc'd with impulsion of blood.

As I said before, the blood that passes through an arterie which is cut and opened, makes no pulse there at all, whence it clearly appears, that the arteries suffer their Diastole neither by innate pulsifick faculty, nor by any granted them from the heart, but by the meer impulsion of the blood. For in the full flux, flowing out the length of its course, you may by touch perceive both the Systole and Diastole, as I said before, and all the differences of the pulse of the heart, their time, order, vehemency, intermission in the emanation of the flux evidently (as it were in a looking glass). Just as water, by the force and impulsion of a spout is driven aloft through pipes of lead, we may observe and distinguish all the forcings of the Engine, though you be a good way off, in the flux of the water when it passes out, the order, beginning, increase, end, and vehemency of every motion. Even so it is when you cut off the orifice of an arterie; where you must observe, That as in the water, the flux is continuall; though it be sometimes nigher, sometimes further; so in the arteries besides the shaking, pulse, and concussion of the blood (which is not equally to be perceived in all), from that time forward there is a continual motion and fluxion in the blood, till the blood be again returned to that place where it first began, that is to say, to the right ear.

These things you may try at your pleasure cutting up one of the longer arteries (as the jugular), which if you take betwixt your fingers, you shall clearly discern how it loses its pulse, and recovers it again, beats lesse or more. And as these things may be tryed whilst the brest is whole; so opening the brest, and the lungs afterwards being collaps'd and all motion of respiration gone, you may easily try it, to wit, that the left ear is contracted and emptyed, that it becomes more whitish, and that it doth at last, together with the left ventricle, intermit in its pulse, beat leisurely, and at last leave off: And likewise by the hole which you may cut in the arterie, you may see the blood come forth lesse and lesse in a smaller thred, and that at last (to wit, in the defeĉt of blood and the impulsion of the left ventricle) no more will flow.

You may likewise try this same in the tying of the *vena arteriosa*, and so take away the pulse of the left ear, and with untying it, restore the pulse at your pleasure. Whence the same thing is evidently tried by experiment, which is seen in dying persons, that as first the left ventricle desists from motion and pulse, and afterwards the left ear, then the right ventricle, and pulse, lastly, the right ear; so where the vital faculty begins first, it ends last.

Which being tried by the sense, it is manifest that the blood passes only through the septum of the heart, and not through the lungs and only through them whilst they are mov'd in respiration, and not when they are fallen or disquieted. For which cause in an Embryon (not as yet breathing) Nature instead of the passage in the *arteria venosa* (that matter may be furnish'd to the left ventricle and the left ear), opens an oval hole which she shuts in young men, and those that breath freely.

It likewise appears, why those that have the vessels of their lungs oppress'd and stuff'd, or those that have any loss of their breath, it is present token of death.

It is likewise clear, why the blood of the lungs is so flame colour'd; for it is thinnest that is strain'd through there. It is beside to be observ'd from our former conclusion in order to those who require the causes of Circulation, and think the power of the heart to be the effecter of all things, and as it is the author of transmission by pulse, so with *Aristotle* they think it the author of attraction, and generation of blood, and that the Spirits are made by the heart and the influxive heat, and that by the innate heat of the heart, as by the immediate instrument of the soul, or by a common bond and the first organ for perfecting of all the works of life. And so the motion of the blood and spirit, its perfection and heat and every property thereof, to be borrow'd from the heart, as from its beginning (which *Aristotle* says is in the blood, as in hot water, or boyling pottage) is in the heart, & that it is the first cause of pulsation and life. If I may speak freely, I do not think that these things are so (as they are commonly believed) for there are many things which perswade me to that opinion, which I will take notice of in the generation of creatures, which are not fit here to be rehearsed; but it may be things more wonderful than these, and such as will give more light to natural Philosophie shall be publish'd by me.

Yet in the mean time I will say and propound it without demonstration (with the leave of most learned men, and reverence to antiquity), that the heart, as it is the beginning of all things in the body, the spring, fountain, and first causer of life, is so to be taken, as being joyn'd together with the veins, and all the arteries, and the blood which is contained in them. Like as the brain (together with all its sensible nerves, organs, and spinal marrow) is the adequate organ of the sense (as the phrase is). But if you understand by this word heart, the body of the heart, with the ventricles and ears, I do not think it to be the framer of the blood, and that it has not force, vertue, motion, or heat, as the gift of the heart; and next, that the same is not the cause of the Diastole and distention, which is the cause of the Systole and contraction, whether in the ears or arteries; but that part of the pulse which is call'd a Diastole, comes of another cause, diverse from the Systole, and ought to go before every Systole. I think the first cause of distention is innate heat in the blood it self, which (like leaven) by little and little attenuated and swelling, is the last thing that is extinct in the creature. I agree to *Aristotle's* instance of pottage, or milk, in so far as he thinks that elevation, or depression of the blood, does not come of vapours, or exhalations, or Spirits rais'd into a vaporous or aereal form, nor is not caus'd by any external agent, but by the regulating of Nature, an internal principle.

Nor is the heart (as some think) like a charcoal-fire, (like a hot Kettle) the beginning of heat and blood, but rather the blood delivers that heat which it has receiv'd to the heart, as likewise to all the rest of the parts, as being the hottest of all. Therefore arteries and the coronal veins are assign'd to the heart for that use which they are assign'd to the rest of the parts, to wit, for influx of heat for the entertaining and conservation of it; therefore all the hotter parts, how much more sanguine they are, and more abundant with blood, they are said convertibly so to be; and thus the heart having signall concavities, is to be thought the Ware-house, continual fire, and fountain of the blood, not because of the corpulency of it, but because of the blood which it contains like a hot Kettle, as in the same manner the spleen, lungs, and other parts are thought hot, because they have many veins or vessels containing blood.

And after this manner do I believe that the native heat call'd innate, to be the first efficient cause of pulse, as likewise to be the common instrument of all operations. This as yet I do not constantly aver, but propound it as a *Thesis*; I would fain know what may be objected by good and learned men, without scurrility of words, reproaches, or base language, and any body shall be welcome to do it.

These things then are as it were the parts, and the footsteps of the passage, and Circulation of the blood; to wit, from the right ear into the ventricle, out of the ventricle through the lungs into the left ear, then into the left ventricle, into the *aorta*, and into all the arteries from the heart, by the porosities of the parts into the veins, and by the veins into the Basis of the heart, the blood returns most speedily.

By an experiment any man may try that pleases by the veins, let the arm be tyed, as the custom is, with a gentle ligature; and let it remain tyed so long,

still moving the arm up and down, till the veins all
of them swell exceedingly, and the skin grow very
red below the ligature, and then let the hand be
washed with Snow or cold water, till the blood
gather'd below the ligature be cold enough, then
presently untying the ligature, you shall find by the
cold blood which returns how swiftly it runs back
to the heart, and what a change it will make in its
return thither; so that it is not to be wondred at,
that in the untying of the ligature in blood-letting
some have sounded. This experiment does demonstrate, that the veins below the ligature do not swell
with blood attenuated, and puft up with spirit, but
with blood only, and such blood which can be reverberated into the arteries through the *Anastomosis*
of the parts, or the hidden Meanders.

It likewise shews how those that pass over snowy
mountains, are often suddenly seiz'd with death,
and many such like.

Lest it should seem a difficult businesse, how the
blood should passe through the pores of the parts,
and go hither and thither, I will add one experiment.
It happens after the same manner to those that are
strangled, and hang'd with a rope, as is does in the
tying of the arm, that beyond the cord, their face,
eyes, lips, tongue, and all the upper parts of their
head are stuff'd with very much blood, grow ex-
tream red, and swell till they look black, in such a
carcase untying the rope, in whatsoever position
you set it, within a very few hours you shall see

all the blood leave the face and the head, and see it as it were fall down with its own weight, from the upper to the lower parts through the pores of the skin and flesh, and the rest of the parts, and that it fills all the parts below, and the skin chiefly, and colours it with black matter; how much more lively and sprightly the blood is in a living body, and by how much more penetrating it is through the porosities than congealed blood, especially when it is condens'd through all the habit of the body, by the cold of death, the ways too being stopp'd and hinder'd, so much the more easy and ready is the passage in those that are alive through all the parts.

Renatus de Cartes, a most acute and ingenious man, (to whom, for his honourable mentioning of my name, I am much indebted) and others with him, when they see the heart of a fish taken out, placed upon an even board, imitate a pulse (by collecting it self) in its erection, up-lifting, vigoration, they think that it is ampliated, and dilated, and that the ventricles of it become more capacious, not according to my opinion. For when it is gathered, at that time the capacities of it are rather streightned, and it is certain that it is then in its Systole, and not in its Diastole, as neither when it falls weak and flagging, and is relax'd, it is then in its Diastole, or distention, and thence the ventricles become wider; so in a dead man, we do not say that his heart is in the Diastole, because it is flagging, without any Systole, destitute of all manner of motion, and not distended at all, for it is distended properly, and is in the Diastole when it is fill'd by the impulsion of the blood, and contraction of the ear, as in the anatomy of living things is evident enough.

Therefore they understand not how much the relaxation, and falling of the heart and arteries differ from their distention and Diastole; that distention, relaxation, and constriction, come not of the same causes, but from contrary causes, as making contrary effects and diverse, as making divers motions; as all Anatomists know very well, that the opposite muscles in any part (called *Antagonistæ*) are the causes of several motions, to wit, of adduction, and extension, so there is necessarily by nature fram'd contrary and divers active organs, for contrary and divers motions.

Nor does this efficient cause of pulse, which he sets down according to *Aristotle*, please me, to wit, that the ebullition of the blood shall be both the cause of the Systole, and of the Diastole. For these motions are sudden strokes, and swift hits. And there is nothing that swells so like leaven, or boyls up so suddenly, in the twinkling of an eye, and falls again; but that rises leisurely, and falls suddenly: besides, in dissection you may by your own eyesight discern, that the ventricles of the heart are distended, and fill'd by the constriction of the ears, and are encreas'd in bignesse according as they are fill'd, more or less, and that the distention of the heart, is a kind of violent motion, done by impulsion, not by an attraction.

There are some who think, as there is no need of impulsion for the aliment in the nourishing of Plants, but it is by little and little drawn into the place of that which is spent by the indigent parts; so the vegetative faculty performs its work alike in both, but there is a difference. Calid influxive is continually requir'd to the entertaining of the members of creatures, and preserving of vivifying heat in them, and for restoring of the parts which suffer by outward injury, and not for nutrition only.

So much of Circulation, which if it be not duly perform'd, or be hinder'd or perverted, or go too swiftly, there follows many dangerous sorts of diseases, and admirable symptoms, either in the veins, as swellings, abscessions, griefs, hæmeroids, flux of blood, or in the arteries, as swellings, boyls, strong and pricking pains, aneurisms, tumors in the flesh, fluxions, sudden suffocations, asthma's, stupidity, apoplexy, and others innumerable. Likewise it is not fit to tell in this place how, as it were with an Enchantment, many things are cur'd, and taken away, which were thought incurable.

I may set down such things in my medicinal observations, and discourses of Pathology, which I have hitherto known to be observ'd by none.

I will conclude (most learned *Riolan*) to give you more ample satisfaction, because you think that there is no Circulation in the mesentericks: Let the *vena porta* be tied near to the *cymus* of the liver in a live dissection, which you may easily try, you shall see by the swelling of the veins beneath the ligature, that same come to pass which happens in blood-letting by tying of the arm, which will shew you the passage of the blood there.

And when you shall hear any man of that opinion, that by *Anastomosis* the blood can come out of the veins into the arteries, tie in a live dissection the great vein, near the division of the crurals, and as soon as you cut the artery (because it finds passage) you shall see all the mass of blood emptied out of all the veins (nay, out of the ascendent *cava* too) by the pulse of the heart, in a very short time, yet that below the ligature the crural veins and parts below are only full. Which, if it could any way have returned into the arteries by an *Anastomosis*, should never have come to passe.

FINIS

Editor's Postscript to the Present Edition
and textual notes

EDITOR'S POSTSCRIPT

TO THE PRESENT EDITION

WILLIAM HARVEY was born in 1578 and died in his eightieth year in 1657. To him it was due more than to any other man that during that period a revolution had taken place in the current ideas about some of the fundamental facts in the physiology of the animal body. For centuries knowledge of the bodily functions had been fettered by a blind belief in the teachings of the great writers such as Aristotle and Galen, and few advances had been made. Beliefs concerning the functions of the heart and blood vessels were so erroneous as now to seem fantastic and absurd, but none had dared to question them until Harvey's master mind set itself to unravel the true facts. His common designation of "discoverer" of the circulation of the blood suggests that he stumbled suddenly and unexpectedly upon the truth. But Harvey, like other scientific investigators, found no short cut to knowledge. His mind had the acuteness necessary to perceive the inconsistencies which had escaped his predecessors, and he possessed the patience and scientific imagination to piece together the innumerable observations that he had made both while treating his patients and while dissecting and experimenting on animals. The result was the announcement that the blood circulated, and this was not a theory, but a correct interpretation of observed facts. Harvey's individuality appeared in his refusal to believe what he was taught. He preferred to follow his own reasoning to its logical conclusion, a conclusion so triumphant that it made his name immortal and has placed the book in which he announced it among the few great books of the world.

Harvey had arrived at his momentous result before he had reached the age of forty. From about 1616 he was explaining his views in his lectures and demonstrations at the Royal College of Physicians, and was no doubt putting them into practice in his wards at St. Bartholomew's Hospital, to which he had been appointed physician in 1609. For twelve years he made nothing public, but eventually in 1628 he allowed his brief Latin treatise entitled *De Motu Cordis* to be printed and published at

Frankfurt-on-Main, then a centre of learning more fitted than London to receive so important an announcement. The consequence was, as he no doubt expected, a storm in the world of philosophy and medecine, which had its reverberations even among the populace of London, so that his reputation and his pocket suffered. Even under severe provocation Harvey did not rush into a controversial defence of his views. His Latin treatise was several times reprinted on the continent of Europe, but he did not deign to publish any further justification of his work until 1649, when a second small book was published at Cambridge and at Rotterdam under the title of *De Circulatione Sanguinis*. This was written in the form of two essays addressed to John Riolan, the younger, a physician and professor of anatomy in the University of Paris.

Harvey's writings were directed solely to the world of science, so that it is not remarkable that they were for long allowed to keep the decent cloak of Latin in which they had first appeared. In 1653, however, when Harvey was old and tired, the two treatises were translated into English and published by Richard Lowndes in London. The writer of the translation remained anonymous. It is unlikely that Harvey himself had any hand in making it, but there is no reason for supposing that he disapproved of the undertaking. The first Latin edition of 1628 had been very inaccurately printed, and later editions repeated many of its mistakes. The translator of 1653 also allowed himself to take liberties in doing his work, and either he or his printer allowed words and even whole sentences to drop right out of the text. This edition was reprinted in 1673, but has never been published again until the present time.

It was not, indeed, until about 1845 that any question of again printing Harvey's writings in English arose. The Sydenham Society then undertook the task and entrusted the preparation of the text to Dr. Robert Willis, a country practitioner. He started with the idea of using the text of 1653, but on examining it was struck with horror. He felt that Harvey's book "must have been rendered into English by one but little conversant with the subject, that it was both extremely rebutting in point of style and full of egregious errors, and that nothing short of an entirely new translation could do justice to this admirable treatise". Willis's translation was published in 1847 and has held the field ever since. Probably no one has ever again troubled to examine the claims of the first translation or to question the justice of Willis's strictures.

It is now 300 years since *De Motu Cordis* was first printed, and a tercentenary edition ought perhaps to reproduce the book in the form in which it first appeared. The world of science is now, however, unable to read Latin with any ease, and the excellence of Harvey's work would not be appreciated in this form. At this distance of time moreover, when the treatise has become a landmark in the history of medecine rather than a text-book, the precise accuracy of a translation has ceased to have any special significance. I have, therefore, returned to the translation of Harvey's own time, and have not been disappointed. The vigour of the seventeenth century is found there with a sufficient sprinkling of expressive, if now unusual, terms to produce the feeling that Harvey himself is speaking. No such illusion is fostered by the dull excellence of the Victorian text, which on a closer inspection, is even found to be not always more accurate than its despised predecessor.

The difference in style is well exemplified by the two translations of the lines of verse in chapter i, the one tersely epigrammatic, the other pompous and verbose. I have not searched the later text with any malicious desire to expose every inaccuracy it may contain, but I have chanced on one or two "egregious errors" in the course of comparing a few passages which I had not perfectly understood. In these instances the text of 1653 turned out to be the more accurate of the two.

There seems, therefore, to be ample justification for returning to the contemporary English text and for presenting Harvey in his own dress to modern eyes.

In preparing the present text the editions of both 1653 and 1673 have been collated. Some small amount of polishing has been necessary. Punctuation has been amended, phrases have occasionally been altered, and a few omissions supplied. Mere printer's errors and faults in punctuation have been silently corrected; all emendations of greater importance will be found recorded in the notes at the end of the volume. I have not thought it necessary to adhere rigidly to every typographical peculiarity of the original text, and I make no apology for inconsistencies in this respect. The compositor of the seventeenth century allowed himself considerable latitude to suit his own convenience, and there is no reason why this should be altogether denied to his successor of the present day.

The original Latin edition of 1628 was adorned with two engraved plates illustrating a series of remarks in the text on the function of the valves in veins. These plates shewed four figures of a man's arm arranged as for blood letting. The hand grips a barber's pole and the upper arm is bound with a tourniquet so as to compress the veins at that level and distend them below. The figures were all adapted from one engraved in a work by Fabricius ab Aquapendente, *De venarum ostiolis*, which had been published in 1604 and again in 1624. The engravings of 1628 were somewhat crude, and were copied with varying degrees of crudity in all the later Latin editions. Unaccountably they were omitted altogether from the English editions of 1653 and 1673, although the references to them were allowed to remain in the text. To remedy this defect a drawing has been made from the life by Stephen Gooden for the present edition and engraved on copper by C. Sigrist; the result is a plate of great beauty, a claim which could not be made for any one of those that have appeared in the editions published hitherto.

GEOFFREY KEYNES.

TEXTUAL NOTES

P. 10, l. 4. The paragraph numbered 3 was omitted in 1653. The translation supplied is my own.

P. 14. l. 8, 9 from bottom, *whey and matter out of the cavity of the breast...*] the Latin text has *è cavitate pectoris serositates & pus Empyicorum...*

P. 20, l. 12, *as if one taking hold*] 1653 has *take* for *taking*, but the Latin shews that the second is right.

P. 20, l. 3 from bottom, *from soft become hard*] 1653 has *of* for *from*.

P. 22, l. 18, *every Fiber circularly plac'd*] for *circularly* 1653 has *circular lyes*.

P. 26, l. 4, *or that they be not at the same instant*] the original text omits *or* and so renders the meaning obscure.

P. 64, last line, *the like*] 1653 has *the life*.

P. 65, l. 8 from bottom, *in cutting*] 1653 has *of cutting*.

P. 68, l. 14, *necessarily*] 1653 has *necessary*.

P. 73, ll. 8—10 from bottom, *we see the testicles to fade and dy, and the great tumours of flesh, and afterwards to fall quite away*] printed in 1653 in a somewhat meaningless form: *the vessels, the testicles, fade and dy, and the great tumours of flesh, and afterwards to fall quite quite away.* Evidently an error has crept in which has here been corrected from the Latin.

P. 76, l. 6, *where it does not flow, they beat not at all*] 1653 omits *at* before *all.* If the Latin original were completely translated it would read: *where it truly does not flow, as in the streight ligature, they beat not at all, exept above the ligature.*

P. 87, ll. 8—3 from bottom, *as if it were... made by the portals*] the translator and printer of 1653 between them rendered these sentences incomprehensible, so that it has been necessary to retranslate them from the Latin.

P. 92, ll. 6, 7, & *elsewhere*] 1653 omits.

P. 111, ll. 6, 5 from bottom, *and in what way... contraho*] omitted in 1653.

P. 112, ll. 6—4 from bottom, *greater than they are in it proportionated, because before the heart is made, or assumes its own function*] this sentence is mistranslated in 1653 as: *greater than proportion, because before the heart is made, that it may do its own function...* The first portion as amended was printed in 1673.

P. 114, l. 10 from bottom, *of the heart, that*] 1653 inserts *and* before *that*.

P. 114, last line, *as a sort of internal animal*] 1653 has *certain* for *sort of*, which is the reading of 1673 and seems to be the better.

P. 126, l. 8 from bottom, *argument and subterfuge*] 1653 omits *and subterfuge*, which is restored in 1673.

P. 127, last line, *and entering into the arteries*] the meaning of these words is not clear, though they are a correct rendering of the Latin text. Dr. Willis overcomes the difficulty in his translation of 1847 by omitting a large part of the passage (*the veins . . . of the carpus*).

P. 137, l. 20, *breaking*] 1653 has *breathing*.

P. 151, l. 16, *constriction*] 1653 has *construction*.

P. 157, l. 12, *one and the same thing*] 1653 omits *one and*.

P. 160, l. 12, *so much I will adjoyn*] 1653 has, after a comma, *only I will adjoyn*, but the Latin text has *tantum addam* after a colon, and this may refer to the next paragraph.

A CATALOG OF SELECTED
DOVER BOOKS
IN SCIENCE AND MATHEMATICS

A CATALOG OF SELECTED

DOVER BOOKS

IN SCIENCE AND MATHEMATICS

QUALITATIVE THEORY OF DIFFERENTIAL EQUATIONS, V.V. Nemytskii and V.V. Stepanov. Classic graduate-level text by two prominent Soviet mathematicians covers classical differential equations as well as topological dynamics and ergodic theory. Bibliographies. 523pp. 5⅜ × 8½. 65954-2 Pa. $10.95

MATRICES AND LINEAR ALGEBRA, Hans Schneider and George Phillip Barker. Basic textbook covers theory of matrices and its applications to systems of linear equations and related topics such as determinants, eigenvalues and differential equations. Numerous exercises. 432pp. 5⅜ × 8½. 66014-1 Pa. $10.95

QUANTUM THEORY, David Bohm. This advanced undergraduate-level text presents the quantum theory in terms of qualitative and imaginative concepts, followed by specific applications worked out in mathematical detail. Preface. Index. 655pp. 5⅜ × 8½. 65969-0 Pa. $14.95

ATOMIC PHYSICS (8th edition), Max Born. Nobel laureate's lucid treatment of kinetic theory of gases, elementary particles, nuclear atom, wave-corpuscles, atomic structure and spectral lines, much more. Over 40 appendices, bibliography. 495pp. 5⅜ × 8½. 65984-4 Pa. $12.95

ELECTRONIC STRUCTURE AND THE PROPERTIES OF SOLIDS: The Physics of the Chemical Bond, Walter A. Harrison. Innovative text offers basic understanding of the electronic structure of covalent and ionic solids, simple metals, transition metals and their compounds. Problems. 1980 edition. 582pp. 6⅛ × 9¼. 66021-4 Pa. $15.95

BOUNDARY VALUE PROBLEMS OF HEAT CONDUCTION, M. Necati Özisik. Systematic, comprehensive treatment of modern mathematical methods of solving problems in heat conduction and diffusion. Numerous examples and problems. Selected references. Appendices. 505pp. 5⅜ × 8½. 65990-9 Pa. $12.95

A SHORT HISTORY OF CHEMISTRY (3rd edition), J.R. Partington. Classic exposition explores origins of chemistry, alchemy, early medical chemistry, nature of atmosphere, theory of valency, laws and structure of atomic theory, much more. 428pp. 5⅜ × 8½. (Available in U.S. only) 65977-1 Pa. $11.95

A HISTORY OF ASTRONOMY, A. Pannekoek. Well-balanced, carefully reasoned study covers such topics as Ptolemaic theory, work of Copernicus, Kepler, Newton, Eddington's work on stars, much more. Illustrated. References. 521pp. 5⅜ × 8½. 65994-1 Pa. $12.95

PRINCIPLES OF METEOROLOGICAL ANALYSIS, Walter J. Saucier. Highly respected, abundantly illustrated classic reviews atmospheric variables, hydrostatics, static stability, various analyses (scalar, cross-section, isobaric, isentropic, more). For intermediate meteorology students. 454pp. 6⅛ × 9¼. 65979-8 Pa. $14.95

RELATIVITY, THERMODYNAMICS AND COSMOLOGY, Richard C. Tolman. Landmark study extends thermodynamics to special, general relativity; also applications of relativistic mechanics, thermodynamics to cosmological models. 501pp. 5⅜ × 8½. 65383-8 Pa. $13.95

APPLIED ANALYSIS, Cornelius Lanczos. Classic work on analysis and design of finite processes for approximating solution of analytical problems. Algebraic equations, matrices, harmonic analysis, quadrature methods, much more. 559pp. 5⅜ × 8½. 65656-X Pa. $13.95

AN INTRODUCTION TO THE PHILOSOPHY OF SCIENCE, Rudolf Carnap. Stimulating, thought-provoking text clearly and discerningly makes accessible such topics as probability, structure of space, causality and determinism, theoretical concepts and much more. 320pp. 5⅜ × 8½. 28318-6 Pa. $8.95

INTRODUCTION TO ANALYSIS, Maxwell Rosenlicht. Unusually clear, accessible coverage of set theory, real number system, metric spaces, continuous functions, Riemann integration, multiple integrals, more. Wide range of problems. Undergraduate level. Bibliography. 254pp. 5⅜ × 8½. 65038-3 Pa. $7.95

INTRODUCTION TO QUANTUM MECHANICS With Applications to Chemistry, Linus Pauling & E. Bright Wilson, Jr. Classic undergraduate text by Nobel Prize winner applies quantum mechanics to chemical and physical problems. Numerous tables and figures enhance the text. Chapter bibliographies. Appendices. Index. 468pp. 5⅜ × 8½. 64871-0 Pa. $11.95

ASYMPTOTIC EXPANSIONS OF INTEGRALS, Norman Bleistein & Richard A. Handelsman. Best introduction to important field with applications in a variety of scientific disciplines. New preface. Problems. Diagrams. Tables. Bibliography. Index. 448pp. 5⅜ × 8½. 65082-0 Pa. $12.95

MATHEMATICS APPLIED TO CONTINUUM MECHANICS, Lee A. Segel. Analyzes models of fluid flow and solid deformation. For upper-level math, science and engineering students. 608pp. 5⅜ × 8½. 65369-2 Pa. $14.95

ELEMENTS OF REAL ANALYSIS, David A. Sprecher. Classic text covers fundamental concepts, real number system, point sets, functions of a real variable, Fourier series, much more. Over 500 exercises. 352pp. 5⅜ × 8½. 65385-4 Pa. $10.95

PHYSICAL PRINCIPLES OF THE QUANTUM THEORY, Werner Heisenberg. Nobel Laureate discusses quantum theory, uncertainty, wave mechanics, work of Dirac, Schroedinger, Compton, Wilson, Einstein, etc. 184pp. 5⅜ × 8½.
60113-7 Pa. $5.95

INTRODUCTORY REAL ANALYSIS, A.N. Kolmogorov, S.V. Fomin. Translated by Richard A. Silverman. Self-contained, evenly paced introduction to real and functional analysis. Some 350 problems. 403pp. 5⅜ × 8½. 61226-0 Pa. $10.95

PROBLEMS AND SOLUTIONS IN QUANTUM CHEMISTRY AND PHYSICS, Charles S. Johnson, Jr. and Lee G. Pedersen. Unusually varied problems, detailed solutions in coverage of quantum mechanics, wave mechanics, angular momentum, molecular spectroscopy, scattering theory, more. 280 problems plus 139 supplementary exercises. 430pp. 6½ × 9¼. 65236-X Pa. $13.95

ASYMPTOTIC METHODS IN ANALYSIS, N.G. de Bruijn. An inexpensive, comprehensive guide to asymptotic methods—the pioneering work that teaches by explaining worked examples in detail. Index. 224pp. 5⅜ × 8½.   64221-6 Pa. $6.95

OPTICAL RESONANCE AND TWO-LEVEL ATOMS, L. Allen and J.H. Eberly. Clear, comprehensive introduction to basic principles behind all quantum optical resonance phenomena. 53 illustrations. Preface. Index. 256pp. 5⅜ × 8½.
65533-4 Pa. $8.95

COMPLEX VARIABLES, Francis J. Flanigan. Unusual approach, delaying complex algebra till harmonic functions have been analyzed from real variable viewpoint. Includes problems with answers. 364pp. 5⅜ × 8½.   61388-7 Pa. $8.95

ATOMIC SPECTRA AND ATOMIC STRUCTURE, Gerhard Herzberg. One of best introductions; especially for specialist in other fields. Treatment is physical rather than mathematical. 80 illustrations. 257pp. 5⅜ × 8½.   60115-3 Pa. $6.95

APPLIED COMPLEX VARIABLES, John W. Dettman. Step-by-step coverage of fundamentals of analytic function theory—plus lucid exposition of five important applications: Potential Theory; Ordinary Differential Equations; Fourier Transforms; Laplace Transforms; Asymptotic Expansions. 66 figures. Exercises at chapter ends. 512pp. 5⅜ × 8½.   64670-X Pa. $12.95

ULTRASONIC ABSORPTION: An Introduction to the Theory of Sound Absorption and Dispersion in Gases, Liquids and Solids, A.B. Bhatia. Standard reference in the field provides a clear, systematically organized introductory review of fundamental concepts for advanced graduate students, research workers. Numerous diagrams. Bibliography. 440pp. 5⅜ × 8½.   64917-2 Pa. $11.95

UNBOUNDED LINEAR OPERATORS: Theory and Applications, Seymour Goldberg. Classic presents systematic treatment of the theory of unbounded linear operators in normed linear spaces with applications to differential equations. Bibliography. 199pp. 5⅜ × 8½.   64830-3 Pa. $7.95

LIGHT SCATTERING BY SMALL PARTICLES, H.C. van de Hulst. Comprehensive treatment including full range of useful approximation methods for researchers in chemistry, meteorology and astronomy. 44 illustrations. 470pp. 5⅜ × 8½.   64228-3 Pa. $11.95

CONFORMAL MAPPING ON RIEMANN SURFACES, Harvey Cohn. Lucid, insightful book presents ideal coverage of subject. 334 exercises make book perfect for self-study. 55 figures. 352pp. 5⅜ × 8¼.   64025-6 Pa. $9.95

OPTICKS, Sir Isaac Newton. Newton's own experiments with spectroscopy, colors, lenses, reflection, refraction, etc., in language the layman can follow. Foreword by Albert Einstein. 532pp. 5⅜ × 8½.   60205-2 Pa. $11.95

GENERALIZED INTEGRAL TRANSFORMATIONS, A.H. Zemanian. Graduate-level study of recent generalizations of the Laplace, Mellin, Hankel, K. Weierstrass, convolution and other simple transformations. Bibliography. 320pp. 5⅜ × 8½.   65375-7 Pa. $8.95

CATALOG OF DOVER BOOKS

THE ELECTROMAGNETIC FIELD, Albert Shadowitz. Comprehensive undergraduate text covers basics of electric and magnetic fields, builds up to electromagnetic theory. Also related topics, including relativity. Over 900 problems. 768pp. 5⅜ × 8¼. 65660-8 Pa. $18.95

FOURIER SERIES, Georgi P. Tolstov. Translated by Richard A. Silverman. A valuable addition to the literature on the subject, moving clearly from subject to subject and theorem to theorem. 107 problems, answers. 336pp. 5⅜ × 8½. 63317-9 Pa. $8.95

THEORY OF ELECTROMAGNETIC WAVE PROPAGATION, Charles Herach Papas. Graduate-level study discusses the Maxwell field equations, radiation from wire antennas, the Doppler effect and more. xiii + 244pp. 5⅜ × 8½. 65678-0 Pa. $6.95

DISTRIBUTION THEORY AND TRANSFORM ANALYSIS: An Introduction to Generalized Functions, with Applications, A.H. Zemanian. Provides basics of distribution theory, describes generalized Fourier and Laplace transformations. Numerous problems. 384pp. 5⅜ × 8½. 65479-6 Pa. $11.95

THE PHYSICS OF WAVES, William C. Elmore and Mark A. Heald. Unique overview of classical wave theory. Acoustics, optics, electromagnetic radiation, more. Ideal as classroom text or for self-study. Problems. 477pp. 5⅜ × 8½. 64926-1 Pa. $12.95

CALCULUS OF VARIATIONS WITH APPLICATIONS, George M. Ewing. Applications-oriented introduction to variational theory develops insight and promotes understanding of specialized books, research papers. Suitable for advanced undergraduate/graduate students as primary, supplementary text. 352pp. 5⅜ × 8½. 64856-7 Pa. $8.95

A TREATISE ON ELECTRICITY AND MAGNETISM, James Clerk Maxwell. Important foundation work of modern physics. Brings to final form Maxwell's theory of electromagnetism and rigorously derives his general equations of field theory. 1,084pp. 5⅜ × 8½. Two-vol. set. Vol. I: 60636-8 Pa. $11.95
Vol. II: 60637-6 Pa. $11.95

AN INTRODUCTION TO THE CALCULUS OF VARIATIONS, Charles Fox. Graduate-level text covers variations of an integral, isoperimetrical problems, least action, special relativity, approximations, more. References. 279pp. 5⅜ × 8½. 65499-0 Pa. $7.95

HYDRODYNAMIC AND HYDROMAGNETIC STABILITY, S. Chandrasekhar. Lucid examination of the Rayleigh-Benard problem; clear coverage of the theory of instabilities causing convection. 704pp. 5⅜ × 8¼. 64071-X Pa. $14.95

CALCULUS OF VARIATIONS, Robert Weinstock. Basic introduction covering isoperimetric problems, theory of elasticity, quantum mechanics, electrostatics, etc. Exercises throughout. 326pp. 5⅜ × 8½. 63069-2 Pa. $8.95

DYNAMICS OF FLUIDS IN POROUS MEDIA, Jacob Bear. For advanced students of ground water hydrology, soil mechanics and physics, drainage and irrigation engineering and more. 335 illustrations. Exercises, with answers. 784pp. 6⅛ × 9¼. 65675-6 Pa. $19.95

CATALOG OF DOVER BOOKS

NUMERICAL METHODS FOR SCIENTISTS AND ENGINEERS, Richard Hamming. Classic text stresses frequency approach in coverage of algorithms, polynomial approximation, Fourier approximation, exponential approximation, other topics. Revised and enlarged 2nd edition. 721pp. 5⅜ × 8½.
65241-6 Pa. $15.95

THEORETICAL SOLID STATE PHYSICS, Vol. I: Perfect Lattices in Equilibrium; Vol. II: Non-Equilibrium and Disorder, William Jones and Norman H. March. Monumental reference work covers fundamental theory of equilibrium properties of perfect crystalline solids, non-equilibrium properties, defects and disordered systems. Appendices. Problems. Preface. Diagrams. Index. Bibliography. Total of 1,301pp. 5⅜ × 8½. Two-vol. set.
Vol. I: 65015-4 Pa. $14.95
Vol. II: 65016-2 Pa. $14.95

OPTIMIZATION THEORY WITH APPLICATIONS, Donald A. Pierre. Broad-spectrum approach to important topic. Classical theory of minima and maxima, calculus of variations, simplex technique and linear programming, more. Many problems, examples. 640pp. 5⅜ × 8½.
65205-X Pa. $14.95

THE CONTINUUM: A Critical Examination of the Foundation of Analysis, Hermann Weyl. Classic of 20th-century foundational research deals with the conceptual problem posed by the continuum. 156pp. 5⅜ × 8½.
67982-9 Pa. $5.95

ESSAYS ON THE THEORY OF NUMBERS, Richard Dedekind. Two classic essays by great German mathematician: on the theory of irrational numbers; and on transfinite numbers and properties of natural numbers. 115pp. 5⅜ × 8½.
21010-3 Pa. $5.95

THE FUNCTIONS OF MATHEMATICAL PHYSICS, Harry Hochstadt. Comprehensive treatment of orthogonal polynomials, hypergeometric functions, Hill's equation, much more. Bibliography. Index. 322pp. 5⅜ × 8½.
65214-9 Pa. $9.95

NUMBER THEORY AND ITS HISTORY, Oystein Ore. Unusually clear, accessible introduction covers counting, properties of numbers, prime numbers, much more. Bibliography. 380pp. 5⅜ × 8½.
65620-9 Pa. $9.95

THE VARIATIONAL PRINCIPLES OF MECHANICS, Cornelius Lanczos. Graduate level coverage of calculus of variations, equations of motion, relativistic mechanics, more. First inexpensive paperbound edition of classic treatise. Index. Bibliography. 418pp. 5⅜ × 8½.
65067-7 Pa. $11.95

MATHEMATICAL TABLES AND FORMULAS, Robert D. Carmichael and Edwin R. Smith. Logarithms, sines, tangents, trig functions, powers, roots, reciprocals, exponential and hyperbolic functions, formulas and theorems. 269pp. 5⅜ × 8½.
60111-0 Pa. $6.95

THEORETICAL PHYSICS, Georg Joos, with Ira M. Freeman. Classic overview covers essential math, mechanics, electromagnetic theory, thermodynamics, quantum mechanics, nuclear physics, other topics. First paperback edition. xxiii + 885pp. 5⅜ × 8½.
65227-0 Pa. $19.95

HANDBOOK OF MATHEMATICAL FUNCTIONS WITH FORMULAS, GRAPHS, AND MATHEMATICAL TABLES, edited by Milton Abramowitz and Irene A. Stegun. Vast compendium: 29 sets of tables, some to as high as 20 places. 1,046pp. 8 × 10½. 61272-4 Pa. $24.95

MATHEMATICAL METHODS IN PHYSICS AND ENGINEERING, John W. Dettman. Algebraically based approach to vectors, mapping, diffraction, other topics in applied math. Also generalized functions, analytic function theory, more. Exercises. 448pp. 5⅜ × 8¼. 65649-7 Pa. $10.95

A SURVEY OF NUMERICAL MATHEMATICS, David M. Young and Robert Todd Gregory. Broad self-contained coverage of computer-oriented numerical algorithms for solving various types of mathematical problems in linear algebra, ordinary and partial, differential equations, much more. Exercises. Total of 1,248pp. 5⅜ × 8½. Two-vol. set. Vol. I: 65691-8 Pa. $14.95
Vol. II: 65692-6 Pa. $14.95

TENSOR ANALYSIS FOR PHYSICISTS, J.A. Schouten. Concise exposition of the mathematical basis of tensor analysis, integrated with well-chosen physical examples of the theory. Exercises. Index. Bibliography. 289pp. 5⅜ × 8½. 65582-2 Pa. $8.95

INTRODUCTION TO NUMERICAL ANALYSIS (2nd Edition), F.B. Hildebrand. Classic, fundamental treatment covers computation, approximation, interpolation, numerical differentiation and integration, other topics. 150 new problems. 669pp. 5⅜ × 8½. 65363-3 Pa. $15.95

INVESTIGATIONS ON THE THEORY OF THE BROWNIAN MOVEMENT, Albert Einstein. Five papers (1905–8) investigating dynamics of Brownian motion and evolving elementary theory. Notes by R. Fürth. 122pp. 5⅜ × 8½. 60304-0 Pa. $4.95

CATASTROPHE THEORY FOR SCIENTISTS AND ENGINEERS, Robert Gilmore. Advanced-level treatment describes mathematics of theory grounded in the work of Poincaré, R. Thom, other mathematicians. Also important applications to problems in mathematics, physics, chemistry and engineering. 1981 edition. References. 28 tables. 397 black-and-white illustrations. xvii + 666pp. 6⅛ × 9¼. 67539-4 Pa. $16.95

AN INTRODUCTION TO STATISTICAL THERMODYNAMICS, Terrell L. Hill. Excellent basic text offers wide-ranging coverage of quantum statistical mechanics, systems of interacting molecules, quantum statistics, more. 523pp. 5⅜ × 8½. 65242-4 Pa. $12.95

ELEMENTARY DIFFERENTIAL EQUATIONS, William Ted Martin and Eric Reissner. Exceptionally clear, comprehensive introduction at undergraduate level. Nature and origin of differential equations, differential equations of first, second and higher orders. Picard's Theorem, much more. Problems with solutions. 331pp. 5⅜ × 8½. 65024-3 Pa. $8.95

STATISTICAL PHYSICS, Gregory H. Wannier. Classic text combines thermodynamics, statistical mechanics and kinetic theory in one unified presentation of thermal physics. Problems with solutions. Bibliography. 532pp. 5⅜ × 8½. 65401-X Pa. $12.95

ORDINARY DIFFERENTIAL EQUATIONS, Morris Tenenbaum and Harry Pollard. Exhaustive survey of ordinary differential equations for undergraduates in mathematics, engineering, science. Thorough analysis of theorems. Diagrams. Bibliography. Index. 818pp. 5⅜ × 8½. 64940-7 Pa. $18.95

STATISTICAL MECHANICS: Principles and Applications, Terrell L. Hill. Standard text covers fundamentals of statistical mechanics, applications to fluctuation theory, imperfect gases, distribution functions, more. 448pp. 5⅜ × 8½. 65390-0 Pa. $11.95

ORDINARY DIFFERENTIAL EQUATIONS AND STABILITY THEORY: An Introduction, David A. Sánchez. Brief, modern treatment. Linear equation, stability theory for autonomous and nonautonomous systems, etc. 164pp. 5⅜ × 8¼. 63828-6 Pa. $5.95

THIRTY YEARS THAT SHOOK PHYSICS: The Story of Quantum Theory, George Gamow. Lucid, accessible introduction to influential theory of energy and matter. Careful explanations of Dirac's anti-particles, Bohr's model of the atom, much more. 12 plates. Numerous drawings. 240pp. 5⅜ × 8½. 24895-X Pa. $6.95

THEORY OF MATRICES, Sam Perlis. Outstanding text covering rank, non-singularity and inverses in connection with the development of canonical matrices under the relation of equivalence, and without the intervention of determinants. Includes exercises. 237pp. 5⅜ × 8½. 66810-X Pa. $7.95

GREAT EXPERIMENTS IN PHYSICS: Firsthand Accounts from Galileo to Einstein, edited by Morris H. Shamos. 25 crucial discoveries: Newton's laws of motion, Chadwick's study of the neutron, Hertz on electromagnetic waves, more. Original accounts clearly annotated. 370pp. 5⅜ × 8½. 25346-5 Pa. $10.95

INTRODUCTION TO PARTIAL DIFFERENTIAL EQUATIONS WITH AP-PLICATIONS, E.C. Zachmanoglou and Dale W. Thoe. Essentials of partial differential equations applied to common problems in engineering and the physical sciences. Problems and answers. 416pp. 5⅜ × 8½. 65251-3 Pa. $10.95

BURNHAM'S CELESTIAL HANDBOOK, Robert Burnham, Jr. Thorough guide to the stars beyond our solar system. Exhaustive treatment. Alphabetical by constellation: Andromeda to Cetus in Vol. 1; Chamaeleon to Orion in Vol. 2; and Pavo to Vulpecula in Vol. 3. Hundreds of illustrations. Index in Vol. 3. 2,000pp. 6⅛ × 9¼. Three-vol. set. Vol. I: 23567-X Pa. $13.95
Vol. II: 23568-8 Pa. $13.95
Vol. III: 23673-0 Pa. $13.95

CHEMICAL MAGIC, Leonard A. Ford. Second Edition, Revised by E. Winston Grundmeier. Over 100 unusual stunts demonstrating cold fire, dust explosions, much more. Text explains scientific principles and stresses safety precautions. 128pp. 5⅜ × 8½. 67628-5 Pa. $5.95

AMATEUR ASTRONOMER'S HANDBOOK, J.B. Sidgwick. Timeless, comprehensive coverage of telescopes, mirrors, lenses, mountings, telescope drives, micrometers, spectroscopes, more. 189 illustrations. 576pp. 5⅜ × 8¼. (Available in U.S. only) 24034-7 Pa. $11.95

SPECIAL FUNCTIONS, N.N. Lebedev. Translated by Richard Silverman. Famous Russian work treating more important special functions, with applications to specific problems of physics and engineering. 38 figures. 308pp. 5⅜ × 8½.
60624-4 Pa. $8.95

OBSERVATIONAL ASTRONOMY FOR AMATEURS, J.B. Sidgwick. Mine of useful data for observation of sun, moon, planets, asteroids, aurorae, meteors, comets, variables, binaries, etc. 39 illustrations. 384pp. 5⅜ × 8¼. (Available in U.S. only)
24033-9 Pa. $8.95

INTEGRAL EQUATIONS, F.G. Tricomi. Authoritative, well-written treatment of extremely useful mathematical tool with wide applications. Volterra Equations, Fredholm Equations, much more. Advanced undergraduate to graduate level. Exercises. Bibliography. 238pp. 5⅜ × 8½.
64828-1 Pa. $7.95

POPULAR LECTURES ON MATHEMATICAL LOGIC, Hao Wang. Noted logician's lucid treatment of historical developments, set theory, model theory, recursion theory and constructivism, proof theory, more. 3 appendixes. Bibliography. 1981 edition. ix + 283pp. 5⅜ × 8½.
67632-3 Pa. $8.95

MODERN NONLINEAR EQUATIONS, Thomas L. Saaty. Emphasizes practical solution of problems; covers seven types of equations. ". . . a welcome contribution to the existing literature. . . ."—Math Reviews. 490pp. 5⅜ × 8½. 64232-1 Pa. $11.95

FUNDAMENTALS OF ASTRODYNAMICS, Roger Bate et al. Modern approach developed by U.S. Air Force Academy. Designed as a first course. Problems, exercises. Numerous illustrations. 455pp. 5⅜ × 8½.
60061-0 Pa. $9.95

INTRODUCTION TO LINEAR ALGEBRA AND DIFFERENTIAL EQUATIONS, John W. Dettman. Excellent text covers complex numbers, determinants, orthonormal bases, Laplace transforms, much more. Exercises with solutions. Undergraduate level. 416pp. 5⅜ × 8½.
65191-6 Pa. $10.95

INCOMPRESSIBLE AERODYNAMICS, edited by Bryan Thwaites. Covers theoretical and experimental treatment of the uniform flow of air and viscous fluids past two-dimensional aerofoils and three-dimensional wings; many other topics. 654pp. 5⅜ × 8½.
65465-6 Pa. $16.95

INTRODUCTION TO DIFFERENCE EQUATIONS, Samuel Goldberg. Exceptionally clear exposition of important discipline with applications to sociology, psychology, economics. Many illustrative examples; over 250 problems. 260pp. 5⅜ × 8½.
65084-7 Pa. $8.95

LAMINAR BOUNDARY LAYERS, edited by L. Rosenhead. Engineering classic covers steady boundary layers in two- and three-dimensional flow, unsteady boundary layers, stability, observational techniques, much more. 708pp. 5⅜ × 8½.
65646-2 Pa. $18.95

LECTURES ON CLASSICAL DIFFERENTIAL GEOMETRY, Second Edition, Dirk J. Struik. Excellent brief introduction covers curves, theory of surfaces, fundamental equations, geometry on a surface, conformal mapping, other topics. Problems. 240pp. 5⅜ × 8½.
65609-8 Pa. $8.95

ROTARY-WING AERODYNAMICS, W.Z. Stepniewski. Clear, concise text covers aerodynamic phenomena of the rotor and offers guidelines for helicopter performance evaluation. Originally prepared for NASA. 537 figures. 640pp. 6⅛ × 9¼.
64647-5 Pa. $15.95

DIFFERENTIAL GEOMETRY, Heinrich W. Guggenheimer. Local differential geometry as an application of advanced calculus and linear algebra. Curvature, transformation groups, surfaces, more. Exercises. 62 figures. 378pp. 5⅜ × 8½.
63433-7 Pa. $8.95

INTRODUCTION TO SPACE DYNAMICS, William Tyrrell Thomson. Comprehensive, classic introduction to space-flight engineering for advanced undergraduate and graduate students. Includes vector algebra, kinematics, transformation of coordinates. Bibliography. Index. 352pp. 5⅜ × 8½.   65113-4 Pa. $8.95

A SURVEY OF MINIMAL SURFACES, Robert Osserman. Up-to-date, in-depth discussion of the field for advanced students. Corrected and enlarged edition covers new developments. Includes numerous problems. 192pp. 5⅜ × 8½.
64998-9 Pa. $8.95

ANALYTICAL MECHANICS OF GEARS, Earle Buckingham. Indispensable reference for modern gear manufacture covers conjugate gear-tooth action, gear-tooth profiles of various gears, many other topics. 263 figures. 102 tables. 546pp. 5⅜ × 8½.   65712-4 Pa. $14.95

SET THEORY AND LOGIC, Robert R. Stoll. Lucid introduction to unified theory of mathematical concepts. Set theory and logic seen as tools for conceptual understanding of real number system. 496pp. 5⅜ × 8¼.   63829-4 Pa. $12.95

A HISTORY OF MECHANICS, René Dugas. Monumental study of mechanical principles from antiquity to quantum mechanics. Contributions of ancient Greeks, Galileo, Leonardo, Kepler, Lagrange, many others. 671pp. 5⅜ × 8½.
65632-2 Pa. $14.95

FAMOUS PROBLEMS OF GEOMETRY AND HOW TO SOLVE THEM, Benjamin Bold. Squaring the circle, trisecting the angle, duplicating the cube: learn their history, why they are impossible to solve, then solve them yourself. 128pp. 5⅜ × 8½.   24297-8 Pa. $4.95

MECHANICAL VIBRATIONS, J.P. Den Hartog. Classic textbook offers lucid explanations and illustrative models, applying theories of vibrations to a variety of practical industrial engineering problems. Numerous figures. 233 problems, solutions. Appendix. Index. Preface. 436pp. 5⅜ × 8½.   64785-4 Pa. $10.95

CURVATURE AND HOMOLOGY, Samuel I. Goldberg. Thorough treatment of specialized branch of differential geometry. Covers Riemannian manifolds, topology of differentiable manifolds, compact Lie groups, other topics. Exercises. 315pp. 5⅜ × 8½.   64314-X Pa. $9.95

HISTORY OF STRENGTH OF MATERIALS, Stephen P. Timoshenko. Excellent historical survey of the strength of materials with many references to the theories of elasticity and structure. 245 figures. 452pp. 5⅜ × 8½. 61187-6 Pa. $11.95

GEOMETRY OF COMPLEX NUMBERS, Hans Schwerdtfeger. Illuminating, widely praised book on analytic geometry of circles, the Moebius transformation, and two-dimensional non-Euclidean geometries. 200pp. 5⅜ × 8¼.
63830-8 Pa. $8.95

MECHANICS, J.P. Den Hartog. A classic introductory text or refresher. Hundreds of applications and design problems illuminate fundamentals of trusses, loaded beams and cables, etc. 334 answered problems. 462pp. 5⅜ × 8½. 60754-2 Pa. $9.95

TOPOLOGY, John G. Hocking and Gail S. Young. Superb one-year course in classical topology. Topological spaces and functions, point-set topology, much more. Examples and problems. Bibliography. Index. 384pp. 5⅜ × 8¼.
65676-4 Pa. $9.95

STRENGTH OF MATERIALS, J.P. Den Hartog. Full, clear treatment of basic material (tension, torsion, bending, etc.) plus advanced material on engineering methods, applications. 350 answered problems. 323pp. 5⅜ × 8½. 60755-0 Pa. $8.95

ELEMENTARY CONCEPTS OF TOPOLOGY, Paul Alexandroff. Elegant, intuitive approach to topology from set-theoretic topology to Betti groups; how concepts of topology are useful in math and physics. 25 figures. 57pp. 5⅜ × 8½.
60747-X Pa. $3.50

ADVANCED STRENGTH OF MATERIALS, J.P. Den Hartog. Superbly written advanced text covers torsion, rotating disks, membrane stresses in shells, much more. Many problems and answers. 388pp. 5⅜ × 8½. 65407-9 Pa. $9.95

COMPUTABILITY AND UNSOLVABILITY, Martin Davis. Classic graduate-level introduction to theory of computability, usually referred to as theory of recurrent functions. New preface and appendix. 288pp. 5⅜ × 8½. 61471-9 Pa. $7.95

GENERAL CHEMISTRY, Linus Pauling. Revised 3rd edition of classic first-year text by Nobel laureate. Atomic and molecular structure, quantum mechanics, statistical mechanics, thermodynamics correlated with descriptive chemistry. Problems. 992pp. 5⅜ × 8½. 65622-5 Pa. $19.95

AN INTRODUCTION TO MATRICES, SETS AND GROUPS FOR SCIENCE STUDENTS, G. Stephenson. Concise, readable text introduces sets, groups, and most importantly, matrices to undergraduate students of physics, chemistry, and engineering. Problems. 164pp. 5⅜ × 8½. 65077-4 Pa. $6.95

THE HISTORICAL BACKGROUND OF CHEMISTRY, Henry M. Leicester. Evolution of ideas, not individual biography. Concentrates on formulation of a coherent set of chemical laws. 260pp. 5⅜ × 8½. 61053-5 Pa. $7.95

THE PHILOSOPHY OF MATHEMATICS: An Introductory Essay, Stephan Körner. Surveys the views of Plato, Aristotle, Leibniz & Kant concerning propositions and theories of applied and pure mathematics. Introduction. Two appendices. Index. 198pp. 5⅜ × 8½. 25048-2 Pa. $7.95

THE DEVELOPMENT OF MODERN CHEMISTRY, Aaron J. Ihde. Authoritative history of chemistry from ancient Greek theory to 20th-century innovation. Covers major chemists and their discoveries. 209 illustrations. 14 tables. Bibliographies. Indices. Appendices. 851pp. 5⅜ × 8½. 64235-6 Pa. $18.95

CATALOG OF DOVER BOOKS

DE RE METALLICA, Georgius Agricola. The famous Hoover translation of greatest treatise on technological chemistry, engineering, geology, mining of early modern times (1556). All 289 original woodcuts. 638pp. 6¾ × 11.
60006-8 Clothbd. $18.95

SOME THEORY OF SAMPLING, William Edwards Deming. Analysis of the problems, theory and design of sampling techniques for social scientists, industrial managers and others who find statistics increasingly important in their work. 61 tables. 90 figures. xvii + 602pp. 5⅜ × 8½. 64684-X Pa. $15.95

THE VARIOUS AND INGENIOUS MACHINES OF AGOSTINO RAMELLI: A Classic Sixteenth-Century Illustrated Treatise on Technology, Agostino Ramelli. One of the most widely known and copied works on machinery in the 16th century. 194 detailed plates of water pumps, grain mills, cranes, more. 608pp. 9 × 12.
28180-9 Pa. $24.95

LINEAR PROGRAMMING AND ECONOMIC ANALYSIS, Robert Dorfman, Paul A. Samuelson and Robert M. Solow. First comprehensive treatment of linear programming in standard economic analysis. Game theory, modern welfare economics, Leontief input-output, more. 525pp. 5⅜ × 8½. 65491-5 Pa. $14.95

ELEMENTARY DECISION THEORY, Herman Chernoff and Lincoln E. Moses. Clear introduction to statistics and statistical theory covers data processing, probability and random variables, testing hypotheses, much more. Exercises. 364pp. 5⅜ × 8½. 65218-1 Pa. $9.95

THE COMPLEAT STRATEGYST: Being a Primer on the Theory of Games of Strategy, J.D. Williams. Highly entertaining classic describes, with many illustrated examples, how to select best strategies in conflict situations. Prefaces. Appendices. 268pp. 5⅜ × 8½. 25101-2 Pa. $7.95

MATHEMATICAL METHODS OF OPERATIONS RESEARCH, Thomas L. Saaty. Classic graduate-level text covers historical background, classical methods of forming models, optimization, game theory, probability, queueing theory, much more. Exercises. Bibliography. 448pp. 5⅜ × 8¼. 65703-5 Pa. $12.95

CONSTRUCTIONS AND COMBINATORIAL PROBLEMS IN DESIGN OF EXPERIMENTS, Damaraju Raghavarao. In-depth reference work examines orthogonal Latin squares, incomplete block designs, tactical configuration, partial geometry, much more. Abundant explanations, examples. 416pp. 5⅜ × 8¼.
65685-3 Pa. $10.95

THE ABSOLUTE DIFFERENTIAL CALCULUS (CALCULUS OF TENSORS), Tullio Levi-Civita. Great 20th-century mathematician's classic work on material necessary for mathematical grasp of theory of relativity. 452pp. 5⅜ × 8½.
63401-9 Pa. $11.95

VECTOR AND TENSOR ANALYSIS WITH APPLICATIONS, A.I. Borisenko and I.E. Tarapov. Concise introduction. Worked-out problems, solutions, exercises. 257pp. 5⅜ × 8¼. 63833-2 Pa. $8.95

THE FOUR-COLOR PROBLEM: Assaults and Conquest, Thomas L. Saaty and Paul G. Kainen. Engrossing, comprehensive account of the century-old combinatorial topological problem, its history and solution. Bibliographies. Index. 110 figures. 228pp. 5⅜ × 8½. 65092-8 Pa. $6.95

CATALYSIS IN CHEMISTRY AND ENZYMOLOGY, William P. Jencks. Exceptionally clear coverage of mechanisms for catalysis, forces in aqueous solution, carbonyl- and acyl-group reactions, practical kinetics, more. 864pp. 5⅜ × 8½. 65460-5 Pa. $19.95

PROBABILITY: An Introduction, Samuel Goldberg. Excellent basic text covers set theory, probability theory for finite sample spaces, binomial theorem, much more. 360 problems. Bibliographies. 322pp. 5⅜ × 8½. 65252-1 Pa. $9.95

LIGHTNING, Martin A. Uman. Revised, updated edition of classic work on the physics of lightning. Phenomena, terminology, measurement, photography, spectroscopy, thunder, more. Reviews recent research. Bibliography. Indices. 320pp. 5⅜ × 8¼. 64575-4 Pa. $8.95

PROBABILITY THEORY: A Concise Course, Y.A. Rozanov. Highly readable, self-contained introduction covers combination of events, dependent events, Bernoulli trials, etc. Translation by Richard Silverman. 148pp. 5⅜ × 8¼.
63544-9 Pa. $6.95

AN INTRODUCTION TO HAMILTONIAN OPTICS, H. A. Buchdahl. Detailed account of the Hamiltonian treatment of aberration theory in geometrical optics. Many classes of optical systems defined in terms of the symmetries they possess. Problems with detailed solutions. 1970 edition. xv + 360pp. 5⅜ × 8½.
67597-1 Pa. $10.95

STATISTICS MANUAL, Edwin L. Crow, et al. Comprehensive, practical collection of classical and modern methods prepared by U.S. Naval Ordnance Test Station. Stress on use. Basics of statistics assumed. 288pp. 5⅜ × 8½.
60599-X Pa. $7.95

DICTIONARY/OUTLINE OF BASIC STATISTICS, John E. Freund and Frank J. Williams. A clear concise dictionary of over 1,000 statistical terms and an outline of statistical formulas covering probability, nonparametric tests, much more. 208pp. 5⅜ × 8½. 66796-0 Pa. $6.95

STATISTICAL METHOD FROM THE VIEWPOINT OF QUALITY CONTROL, Walter A. Shewhart. Important text explains regulation of variables, uses of statistical control to achieve quality control in industry, agriculture, other areas. 192pp. 5⅜ × 8½. 65232-7 Pa. $7.95

THE INTERPRETATION OF GEOLOGICAL PHASE DIAGRAMS, Ernest G. Ehlers. Clear, concise text emphasizes diagrams of systems under fluid or containing pressure; also coverage of complex binary systems, hydrothermal melting, more. 288pp. 6½ × 9¼. 65389-7 Pa. $10.95

STATISTICAL ADJUSTMENT OF DATA, W. Edwards Deming. Introduction to basic concepts of statistics, curve fitting, least squares solution, conditions without parameter, conditions containing parameters. 26 exercises worked out. 271pp. 5⅜ × 8½. 64685-8 Pa. $8.95

TENSOR CALCULUS, J.L. Synge and A. Schild. Widely used introductory text covers spaces and tensors, basic operations in Riemannian space, non-Riemannian spaces, etc. 324pp. 5⅜ × 8¼. 63612-7 Pa. $8.95

A CONCISE HISTORY OF MATHEMATICS, Dirk J. Struik. The best brief history of mathematics. Stresses origins and covers every major figure from ancient Near East to 19th century. 41 illustrations. 195pp. 5⅜ × 8½. 60255-9 Pa. $7.95

A SHORT ACCOUNT OF THE HISTORY OF MATHEMATICS, W.W. Rouse Ball. One of clearest, most authoritative surveys from the Egyptians and Phoenicians through 19th-century figures such as Grassman, Galois, Riemann. Fourth edition. 522pp. 5⅜ × 8½. 20630-0 Pa. $11.95

HISTORY OF MATHEMATICS, David E. Smith. Nontechnical survey from ancient Greece and Orient to late 19th century; evolution of arithmetic, geometry, trigonometry, calculating devices, algebra, the calculus. 362 illustrations. 1,355pp. 5⅜ × 8½. Two-vol. set. Vol. I: 20429-4 Pa. $12.95
Vol. II: 20430-8 Pa. $11.95

THE GEOMETRY OF RENÉ DESCARTES, René Descartes. The great work founded analytical geometry. Original French text, Descartes' own diagrams, together with definitive Smith-Latham translation. 244pp. 5⅜ × 8½. 60068-8 Pa. $7.95

THE ORIGINS OF THE INFINITESIMAL CALCULUS, Margaret E. Baron. Only fully detailed and documented account of crucial discipline: origins; development by Galileo, Kepler, Cavalieri; contributions of Newton, Leibniz, more. 304pp. 5⅜ × 8½. (Available in U.S. and Canada only) 65371-4 Pa. $9.95

THE HISTORY OF THE CALCULUS AND ITS CONCEPTUAL DEVELOPMENT, Carl B. Boyer. Origins in antiquity, medieval contributions, work of Newton, Leibniz, rigorous formulation. Treatment is verbal. 346pp. 5⅜ × 8½. 60509-4 Pa. $9.95

THE THIRTEEN BOOKS OF EUCLID'S ELEMENTS, translated with introduction and commentary by Sir Thomas L. Heath. Definitive edition. Textual and linguistic notes, mathematical analysis. 2,500 years of critical commentary. Not abridged. 1,414pp. 5⅜ × 8½. Three-vol. set. Vol. I: 60088-2 Pa. $9.95
Vol. II: 60089-0 Pa. $9.95
Vol. III: 60090-4 Pa. $9.95

GAMES AND DECISIONS: Introduction and Critical Survey, R. Duncan Luce and Howard Raiffa. Superb nontechnical introduction to game theory, primarily applied to social sciences. Utility theory, zero-sum games, n-person games, decision-making, much more. Bibliography. 509pp. 5⅜ × 8½. 65943-7 Pa. $12.95

THE HISTORICAL ROOTS OF ELEMENTARY MATHEMATICS, Lucas N.H. Bunt, Phillip S. Jones, and Jack D. Bedient. Fundamental underpinnings of modern arithmetic, algebra, geometry and number systems derived from ancient civilizations. 320pp. 5⅜ × 8½. 25563-8 Pa. $8.95

CALCULUS REFRESHER FOR TECHNICAL PEOPLE, A. Albert Klaf. Covers important aspects of integral and differential calculus via 756 questions. 566 problems, most answered. 431pp. 5⅜ × 8½. 20370-0 Pa. $8.95

CATALOG OF DOVER BOOKS

CHALLENGING MATHEMATICAL PROBLEMS WITH ELEMENTARY SOLUTIONS, A.M. Yaglom and I.M. Yaglom. Over 170 challenging problems on probability theory, combinatorial analysis, points and lines, topology, convex polygons, many other topics. Solutions. Total of 445pp. 5⅜ × 8½. Two-vol. set.
Vol. I 65536-9 Pa. $7.95
Vol. II 65537-7 Pa. $6.95

FIFTY CHALLENGING PROBLEMS IN PROBABILITY WITH SOLUTIONS, Frederick Mosteller. Remarkable puzzlers, graded in difficulty, illustrate elementary and advanced aspects of probability. Detailed solutions. 88pp. 5⅜ × 8½.
65355-2 Pa. $4.95

EXPERIMENTS IN TOPOLOGY, Stephen Barr. Classic, lively explanation of one of the byways of mathematics. Klein bottles, Moebius strips, projective planes, map coloring, problem of the Koenigsberg bridges, much more, described with clarity and wit. 43 figures. 210pp. 5⅜ × 8½. 25933-1 Pa. $6.95

RELATIVITY IN ILLUSTRATIONS, Jacob T. Schwartz. Clear nontechnical treatment makes relativity more accessible than ever before. Over 60 drawings illustrate concepts more clearly than text alone. Only high school geometry needed. Bibliography. 128pp. 6⅛ × 9¼. 25965-X Pa. $7.95

AN INTRODUCTION TO ORDINARY DIFFERENTIAL EQUATIONS, Earl A. Coddington. A thorough and systematic first course in elementary differential equations for undergraduates in mathematics and science, with many exercises and problems (with answers). Index. 304pp. 5⅜ × 8½. 65942-9 Pa. $8.95

FOURIER SERIES AND ORTHOGONAL FUNCTIONS, Harry F. Davis. An incisive text combining theory and practical example to introduce Fourier series, orthogonal functions and applications of the Fourier method to boundary-value problems. 570 exercises. Answers and notes. 416pp. 5⅜ × 8½. 65973-9 Pa. $11.95

THE THEORY OF BRANCHING PROCESSES, Theodore E. Harris. First systematic, comprehensive treatment of branching (i.e. multiplicative) processes and their applications. Galton-Watson model, Markov branching processes, electron-photon cascade, many other topics. Rigorous proofs. Bibliography. 240pp. 5⅜ × 8½. 65952-6 Pa. $6.95

AN INTRODUCTION TO ALGEBRAIC STRUCTURES, Joseph Landin. Superb self-contained text covers "abstract algebra": sets and numbers, theory of groups, theory of rings, much more. Numerous well-chosen examples, exercises. 247pp. 5⅜ × 8½. 65940-2 Pa. $7.95

Prices subject to change without notice.

Available at your book dealer or write for free Mathematics and Science Catalog to Dept. GI, Dover Publications, Inc., 31 East 2nd St., Mineola, N.Y. 11501. Dover publishes more than 250 books each year on science, elementary and advanced mathematics, biology, music, art, literature, history, social sciences and other areas.